Yale Agrarian Studies Series

James C. Scott, *series editor*

The Agrarian Studies Series at Yale University Press seeks to publish outstanding and original interdisciplinary work on agriculture and rural society—for any period, in any location. Works of daring that question existing paradigms and fill abstract categories with the lived experience of rural people are especially encouraged.

—James C. Scott, *Series Editor*

James C. Scott, *Seeing Like a State: How Certain Schemes to Improve the Human Condition Have Failed*

Steve Striffler, *Chicken: The Dangerous Transformation of America's Favorite Food*

Alissa Hamilton, *Squeezed: What You Don't Know About Orange Juice*

James C. Scott, *The Art of Not Being Governed: An Anarchist History of Upland Southeast Asia*

Michael R. Dove, *The Banana Tree at the Gate: A History of Marginal Peoples and Global Markets in Borneo*

Edwin C. Hagenstein, Sara M. Gregg, and Brian Donahue, eds., *American Georgics: Writings on Farming, Culture, and the Land*

Timothy Pachirat, *Every Twelve Seconds: Industrialized Slaughter and the Politics of Sight*

Andrew Sluyter, *Black Ranching Frontiers: African Cattle Herders of the Atlantic World, 1500–1900*

Brian Gareau, *From Precaution to Profit: Contemporary Challenges to Environmental Protection in the Montreal Protocol*

Kuntala Lahiri-Dutt and Gopa Samanta, *Dancing with the River: People and Life on the Chars of South Asia*

Alon Tal, *All the Trees of the Forest: Israel's Woodlands from the Bible to the Present*

Felix Wemheuer, *Famine Politics in Maoist China and the Soviet Union*

Jenny Leigh Smith, *Works in Progress: Plans and Realities on Soviet Farms, 1930–1963*

Graeme Auld, *Constructing Private Governance: The Rise and Evolution of Forest, Coffee, and Fisheries Certification*

Jess Gilbert, *Planning Democracy: Agrarian Intellectuals and the Intended New Deal*

Jessica Barnes and Michael R. Dove, eds., *Climate Cultures: Anthropological Perspectives on Climate Change*

Shafqat Hussain, *Remoteness and Modernity: Transformation and Continuity in Northern Pakistan*

For a complete list of titles in the Yale Agrarian Studies Series, visit yalebooks.com/agrarian.

Climate Cultures

Anthropological Perspectives on Climate Change

Edited by Jessica Barnes and Michael R. Dove

Yale UNIVERSITY PRESS
NEW HAVEN & LONDON

Cover image: Ice sculptures by the Brazilian artist Néle Azevedo in Gendarmenmarkt public square in Berlin, September 2, 2009. REUTERS/Tobias Schwarz.

Yale University Press books may be purchased in quantity for educational, business, or promotional use. For information, please e-mail sales.press@yale.edu (U.S. office) or sales@yaleup.co.uk (U.K. office).

Set in Adobe Garamond type by IDS Infotech, Ltd.
Printed in the United States of America.

ISBN: 978-0-300-19881-2 (paperback)

Library of Congress Control Number: 2014958688

A catalogue record for this book is available from the British Library.

This paper meets the requirements of ANSI/NISO Z39.48–1992 (Permanence of Paper).

10 9 8 7 6 5 4 3 2 1

Contents

Part II. Knowing Climate Change

Part III. Imagining Climate Change

Preface

Climate change has become one of the most pressing issues of our time. Increasing air and ocean temperatures, altered precipitation and storm patterns, and rising sea levels are affecting the globe, with profound social, political, and economic consequences. Academia has responded to this challenge with a proliferation in the number of disciplines and scholars pursuing research related to climate change. These scholars include many anthropologists who are thinking about the nexus of nature, culture, science, politics, and belief that constitutes climate change. Yet although conceptually anthropology has much to offer the field, there are few books that represent the range of anthropological perspectives on climate change. The goal of this book is to fill that gap, by integrating and articulating the contributions that anthropology can bring to our understandings of and response to climate change.

In the introduction we identify the key ways in which we see anthropological scholarship as adding to climate change debates. The main part of the book is organized in three parts. The first historicizes climate change, drawing important parallels between present-day discourse and practice and what has gone before. The second explores how the body of knowledge about climate change and its impacts is produced and interpreted by scientists, modelers, policymakers, and community members. The third highlights the role of the imagination in shaping conceptions of climate change. Given its broad range of case studies from diverse time periods and geographic areas, the book will be of interest to scholars in anthropology, geography, sociology, environmental history, political ecology, political science, environmental studies, science and technology studies, and interdisciplinary programs on climate change or, more broadly, society–environment relations. Climate change often appears to be an intractable problem, and certainly some dimensions of it may prove to be so. Yet anthropology offers new possibilities for unpacking the complex relationships between society and climate, certainty and uncertainty, science and knowledge, global and local that embody climate change. To those who participate in and to those who simply observe the polarized public debates about climate change, we offer nuanced and grounded ways of thinking about this challenge.

Publication of this book was made possible by a number of institutions and individuals. We thank the funders of a workshop that both energized and helped to frame the development of this book: the Yale Climate and Energy Institute, Yale School of Forestry and Environmental Studies, Edward J. and Dorothy Clarke Kempf Memorial Fund at the MacMillan Center for International and Area Studies at Yale, Yale Tropical Resources Institute, and Yale Institute of Sacred Music. We thank our current institutions, the University of South Carolina and Yale School of Forestry and Environmental Studies, for their support. We thank Shereen D'Souza, who helped coordinate the workshop, and Kelly Goldberg for her help with the preparation of the manuscript. We thank Bill Nelson and Lynn Shirley for their assistance in preparing the figures for the text and Alexander Trotter for writing the index. At Yale University Press, we especially thank Jean Thomson Black for her backing of this project from the outset, including attending our workshop, Margaret Otzel and Samantha Ostrowski.

Introduction

Jessica Barnes and Michael R. Dove

A recent summer edition of the *New Yorker* magazine adopted an unseasonal topic for its front cover: Santa Claus (fig. 0.1).[1] In the illustration, Santa, his cheeks flushed, is slumped on the ground against a striped pole, under the yellow orb of a bright sun. His expression is one of dismay, surprise, or perhaps exhaustion. His fairy-tale frozen homeland has been reduced to a small iceberg in a sea of blue. There is little room for the elves to work or the reindeer to eat here. Santa is marooned.

The magazine cover conveys a powerful message of environmental changes that are happening today. The centrally placed sun evokes a warmer, climate-changed world. The isolated iceberg captures a landscape fundamentally altered by that sun. The figure of Santa Claus strikes a note of familiarity, yet his slouched stance and sad expression suggest that something is amiss. At the same time, the presence of this legendary figure in Western folklore alludes to ongoing questions being asked by many members of the North American public about whether these environmental changes are in fact real or, like Santa, just a myth. Despite an assembled

0.1: Santa on the cover of *The New Yorker*. Ian Falconer for *The New Yorker*. Courtesy of *The New Yorker*.

body of evidence that links increasing air and ocean temperatures, altered precipitation and storm patterns, and rising sea levels to higher concentrations of greenhouse gases in the atmosphere (IPCC 2013), climate change continues to arouse skepticism and elude comprehension.

The central proposition of this book is that anthropology can offer a valuable set of insights into the nexus of nature, culture, science, politics, and belief that constitutes climate change. As the *New Yorker* cover implies, climate change is not just about hotter temperatures and melting ice. It is also about stories and images, myth and reality, knowledge and ignorance, humor and tragedy—questions that are, at root, cultural in nature. While this collision and confusion of forces flummox climate scientists and policy makers alike, in some ways they make it all the more attractive to the anthropological project (Batterbury 2008; Crate 2011; Crate and Nuttall 2009; Diemberger et al. 2012; Lahsen 2007; Strauss and Orlove 2003; Townsend 2010). Indeed the intersection of so many disparate phenomena was long ago claimed by anthropology as a domain not only in which it could operate but also in which it excels. Furthermore, just as anthropology speaks to climate change, so, too, does climate change speak to broader questions within the discipline on the relationship between society and the environment, state and citizens, science and knowledge, global and local.

Climate change can be approached from many perspectives: from that of the climate modeling work that is done to parcel out the signature of anthropogenic climate change from other drivers of climatic variability; or the studies that probe the impacts of changing climatic patterns on national economies or plant and animal habitats; or the analyses of the factors that compel some politicians to take action on reducing greenhouse gas emissions and others not. But anthropology offers a different perspective. Guided by this discipline's unique set of theoretical and methodological frameworks, an anthropologist might ask: What forms of scientific expertise are employed to determine that climate change is taking place? How is the science of climate change culturally and politically mediated? Who participates in the production of knowledge about global climate change, and how does this participation shape the reception of its findings by diverse groups? How do postcolonial and North–South politics affect national and international stances on climate change? What are the terms in which new climatic patterns are conceptualized, challenged, and negotiated in disparate social worlds? How is ignorance as well as knowledge about climate change constructed and circulated? How do programs for climate change adaptation and mitigation rework deeply embedded and culturally constituted human interactions with their environments?

The goal of this book is to identify, integrate, and articulate the contributions that anthropological perspectives can bring to our understandings of and responses to climate change. In doing so, it builds on recent attempts by a number of anthropologists to communicate to a broader audience the value of anthropological insights on climate change (Barnes et al. 2013; Hastrup 2013). This is an opportune moment. At a time of mounting challenges to action on climate change, as international negotiations flounder and climate skeptics capture the attention of the public, there is an opening for greater voice from the social sciences (Agrawal et al. 2012; Hulme 2011a; Jasanoff 2010; Vinthagen 2013; Yearley 2009). An initial assumption among many within the climate science research community that their findings alone would be sufficient to spur action on climate change has been shown to be unfounded, given the unexpected and resolute mistrust of this science in many segments of the public, especially in the United States. As a result, many climate scientists have an increasing interest in what social scientists have to say. This is a book, therefore, both for anthropologists interested in understanding more about how societies engage with transforming environments, and for scholars from multiple other disciplines interested in exploring new ways of understanding the challenge that anthropogenic climate change poses to contemporary society.

Presenting chapters that draw on case studies from a range of time periods and geographic areas, the book explores the historical context of contemporary climate change debates, the types of knowledge various actors draw on to interpret climate change, and the role of the imagination in shaping understandings of climate–society interactions.

ETHNOGRAPHY

Anthropological perspectives on climate change are shaped by ethnography—the methodology that in many ways defines anthropology as a discipline. Ethnography is the study and "thick description" (Geertz 1973) of people's behavior and social relations in the cultural and ecological contexts in which they live. Through immersion in people's social worlds, ethnographers gain an insight into the quotidian interactions between people and their environments. Ethnographic research opens a window into how people's everyday activities, relationships, and value systems both affect and are affected by changing environmental conditions. From native groups in the Arctic (Crate 2008; Hastrup 2012; Henshaw 2009; Marino and Schweitzer 2009; Nuttall 2009) to coastal residents in Papua New Guinea (Lipset 2011), islanders in the Pacific (Rudiak-Gould 2013), rice cultivators in West Africa (Davidson 2013), and farmers in the Andes (Orlove et al. 2002), people are observing and responding to ongoing changes in their weather, climate, and landscapes. These detailed, on-the-ground outlooks are a valuable contribution to a debate that is often dominated by discussions of global atmospheric processes and international policy negotiations (Magistro and Roncoli 2001). It is important to think about how people in particular contexts are perceiving and adapting to changes linked to climate change. It is also important to think about how people in other contexts may *not* be concerned about climate change (Barnes, this volume). Perceptions of and actions on climate change are shaped by sociocultural relations of power, such as gender (Bee 2013), and by the ways in which people see themselves in the world (Rudiak-Gould 2013). We can learn a great deal by looking at what people do as they engage with their environments day by day and how they talk about what they do (Puri, this volume).

While other disciplines often associate anthropology with providing an insight into "the local," this is not the discipline's only potential contribution to climate change debates (Lahsen 2007). Indeed, over the past few decades many anthropologists have been moving away from studies of individual communities to analyses of the ways in which people, objects, and ideas are interrelated across space and time in an increasingly globalized world (Gupta and Ferguson 1997; Marcus 1995; Tsing 2005). This shift has led anthropologists to draw connections between seemingly isolated local places and wider national and global politics. It has also meant expanding the topics of anthropological study to include research settings in developed countries and institutional centers of power, and diverse research subjects ranging from nongovernmental organizations to policy makers, scientists, activists, international agencies, and corporations. The chapters in this book hence discuss many actors, from indigenous communities to scientists, government bureaucrats to computer modelers, philosophers to international development practitioners.

Illustrative of this shifting emphasis within the discipline is the growing body of ethnographic work on climate scientists and the production of climate change knowledge. Building on scholarship in science and technology studies on scientific practice, the circulation of scientific knowledge, and the relationship between science and politics (Callon et al. 1986; Collins and Pinch 1982; Jasanoff 2004; Knorr Cetina 1999; Latour and Woolgar 1986; Pickering 1995), this work seeks not to critique or undermine the value of scientific knowledge but to ask how we know what we know about climate change. Through ethnographic studies of climate scientists and international scientific meetings, anthropologists reveal how science is generated within particular social, political, economic, and cultural contexts—contexts that influence the research questions asked, the methods selected to answer those questions, and, ultimately, the results generated (Günel 2012; Hastrup and Skrydstrup 2013; Lahsen 2005; Lahsen, this volume; Moore 2012; O'Reilly et al. 2012; O'Reilly, this volume).

Yet what we know about climate change is a matter not only of the knowledge that is produced but also of how that knowledge moves through different spaces and is interpreted and reinterpreted along the way. Ethnographic enquiry can provide important insights into how the science of climate change circulates through everyday practice, policy realms, media discourse, activist movements, and popular culture (Diemberger et al. 2012). This circulation is contingent on translation from scientific to vernacular knowledge, a process in which the understandings of key terms such as *climate* can shift, as words assume new meanings in different languages and cultures (Rudiak-Gould 2012). Political dynamics and cultural interpretations strongly influence how and where these scientific facts circulate. Some facts come to be politically contested, as countries negotiate over assigning responsibility for the causation and mitigation of climate change, while others are not (Dove 1994; Mahony 2014). Some facts are taken onboard by policy makers while others are rejected owing to the relations of trust between decision makers and national and international scientists, and the geopolitical underpinnings this knowledge acquires in specific contexts (Lahsen 2009). These dynamics influence both how people in different places see the science of climate change and how that science, in turn, makes people see those places in different ways (Moore 2010).

Many people working on climate change pigeonhole anthropology's remit—within the commonly framed triad of impacts, adaptation, and mitigation—as adaptation. By offering an understanding of social practice, cultural perceptions, power relations, and societal resilience, an anthropological lens can indeed shed light on how communities are already adapting to changing climatic conditions and how future adaptation initiatives may play out (Brondizio and Moran 2008; Hastrup 2009; McIntosh, this volume;

Orlove 2009; Puri, this volume; Thornton and Manasfi 2010; Yager, this volume). But ethnographic inquiry also informs other dimensions of climate change. It helps us understand the science of climate change as being socially and culturally mediated (Lahsen 2002; Lahsen, this volume; O'Reilly, this volume). It elucidates how certain climate change impacts come to be prioritized over others (Orlove et al., this volume). And it reveals how strategies to mitigate climate change must be seen in a larger context of state efforts to control diverse groups' interactions with their resource base (Lansing 2011; McElwee, this volume; Mathews, this volume).

HISTORY

Anthropology offers a valuable historical contextualization of contemporary climate change. Anthropologists have long been interested in the social patterns and processes that emerge through time as people interact with their environments and with each other. One of the central things archaeologists look at, for example, is how past peoples forged their livelihoods and related to one another in the face of considerable natural fluctuations in climate (Anderson et al. 2006). Archaeological investigation has documented societal collapse and settlement abandonment as an adaptive mechanism in the face of decadal- or century-scale droughts (Axtell et al. 2002; Bar-Yosef 2011; Dillehay and Kolata 2004; Kaniewski et al. 2012; Weiss et al. 1993). In other cases, archaeologists have found evidence of societies' abilities to accommodate climatic change with less radical adaptations (Parker 2013). This work indicates the diversity of adaptive mechanisms that may be drawn upon in response to future climate change (McIntosh et al. 2000; McIntosh, this volume; Orlove 2005; Rosen 2007; Weiss and Bradley 2001).

Just as significant as the history of human experience with climatic change is the history of human thinking about society–climate relations. Although the anthropogenic forcing of global climate through greenhouse gas emissions over the past century is unprecedented, many of the questions it raises have their roots in much older discussions about society–environment interactions. From the time of Hippocrates and before, scholars have asked questions about the degree to which climates determine societal characteristics and have pondered the limits to human agency in its ability to manage the environment (Dove, this volume). For millennia there have been debates about why one's place on this earth is different from another's—historical precedents to current conversations about climate change impacting some places more than others (Orlove et al., this volume). In the modern era, communities have developed contrasting belief systems regarding how landcover change might alter the climate and have formulated land-use policies and practices accordingly (Mathews 2009; Mathews, this volume). The science of understanding climate change as a discursive phenomenon, too, has an important history (Fleming 1998; Hamblyn 2009; Sorlin 2009). This historical perspective is a useful reminder that environments have never been static and people have always not only affected their environments but also have been affected by them and have perceived, interpreted, and responded to these changes (McIntosh, this volume).

Furthermore, anthropologists not only attend to history—both the history of human experience with climate and the history of human thinking about climate—but also consider the very idea of history (Sahlins 2004; Wolf 1982). This opens the door to studying the way in which ideas of history shape contemporary thinking about climate change (Dove, this volume). It also allows anthropologists to problematize, for example, the proposition that anthropogenic climate change marks the "end of history" (Chakrabarty 2009).

The final dimension of the discipline's historical perspective is the way in which anthropologists see climate change as but the latest in a series of global concerns that have occupied the attention of the international community over recent decades. These issues include, for instance, concerns about international development, biodiversity conservation, and protected area management. Indeed there are a number of parallels between climate change and these other issues that generate similar sorts of literatures, international debate, and policy action. Much of the discussion about how climate change will affect vulnerable communities and the need for adaptation, for example, resonates with earlier discussions about the uneven pace of development. The developed countries have contributed most to the emissions that drive climate change, and yet the developing countries will bear most of the burden of impacts, just as some argue that the political economies of colonialism and postcolonial trade relations have benefited the wealthy regions at the expense of the poor (Bunker 1985; Cockcroft et al. 1972; Frank 1967). Climate change *adaptation* has become the new development buzzword. It has become the hot topic of the moment for researchers and program directors who seek international financial support, the successor to terms like *basic needs, participation,* and *integrated conservation and development,* which led earlier waves of development intervention and funding flows to developing countries. In some cases, international development practitioners are simply replacing the label *underdeveloped* with the label *low adaptive capacity* in assistance programs, while paying little attention to the dissimilarities involved (Moore 2012). Many local communities, too, are seeing how framing their development challenges in terms of climate change, and the need to address those challenges in terms of climate change adaptation, can facilitate access to new social networks, spaces of action, and funding streams (Jonsdottir 2013).

Similarly, current efforts to mitigate climate change by reducing emissions from deforestation and degradation (REDD) can be viewed from the perspective of a long history of efforts to change the ways in which people in the tropics use their land, whether through preserving forest cover in the name of biodiversity (McElwee, this volume) or adopting particular forms of agricultural production in the name of economic development and environmental sustainability (Lansing 2011). Just as past development projects that cast forest-dwelling communities as the source of degradation and ignored the role of timber companies and parastatal plantations met with little success (McElwee 2010), efforts to mitigate climate change by paying less developed countries to maintain standing forest cover are unlikely to succeed unless they address the actual factors underlying deforestation, including rising demand in more developed countries for industrial crops and products like soybeans, palm oil, biofuels, and beef.

A HOLISTIC VIEW

The third contribution of an anthropological perspective is its holistic view of society and the environment, which rejects a narrow focus on one variable or set of variables and instead strives to recognize all possible determinants of social and environmental change.[2] Such a holistic view situates climate change within a broader set of contextual relations and highlights the fact that climate change can only ever be one of a number of influences on people's social lives. It draws attention to the fact that the new forms of production and consumption driving contemporary climate change are at the same time altering people's livelihood strategies, modes of interaction, and spatial and temporal horizons (Crate 2011; Ferguson 2006; Tsing 2005). Hence, climate change is inevitably accompanied everywhere by other kinds of changes within society (Yager, this volume). While climate is sometimes the dominant factor driving change at a given time and place, it is often outweighed by other factors (Barnes, this volume).

A holistic approach can help unpack some of the issues of scale involved in studying climate change. At larger spatial and temporal scales, the "fingerprint" of anthropogenic climate change is easy to identify, and predictions of global temperature increase can be made with a fair level of certainty. But at the smaller scales at which everyday lives are affected and policy is implemented, it is harder for climate scientists to project changes and their outcomes. Attributing events and trends at this level to climate change is thus far more difficult (Moore et al., this volume). At the local scale, the hegemony of climatic explanation is displaced by a hegemony of social explanation. As climate science becomes silent, advocates of other views become more voluble. In some places, people are talking and worrying about climate change, but in many places they are not (Barnes, this volume).

Yet despite the fact that climate change is but one of a number of ongoing dynamics that shape societies, it is becoming ever more prominent an explanation for a wide range of social and economic issues, from crop failure to transborder refugees to issues of national and international security (Fleming and Jankovic 2011). With increasing scholarly and political attention focused on climate change, there is a growing tendency to ascribe all changes in environment and society to climate—a pattern that Hulme (2011b) describes as "climate reductionism." Here, anthropology can play a crucial role by situating the repercussions of climate change in the context of other processes that may actually have more immediate significance to people's lives, such as grinding poverty or the loss of arable land and biodiversity. Climate change cannot be unraveled from the complex web of social and material relations that mediate people's interactions with their environments (Cassidy 2012). Nor can we understand climate change without relating it to the industrial, market-oriented systems of production that produce greenhouse gases and make emission reductions so politically and economically intractable today.

Isolating climate change from other variables can serve powerful discursive purposes for drawing attention to the issue of climate change and its impacts, as in the case of "climate refugees" (Bettini 2013). But it can also depoliticize the processes at work and undermine the agency of those concerned. Whereas the media portray potential climate refugees as passive victims of a changing climate, for example, a more holistic view reveals that for many of these people migration has historically been a central part of their everyday lives, with or without climate change (Farbotko and Lazrus 2012).

OVERVIEW OF THE BOOK

Part I. Historicizing Climate Change

The first part of the book historicizes climate change. Although many scholars and popular commentators frame contemporary climate change as an unprecedented challenge—a break with history (Chakrabarty 2009)—the first three chapters illustrate some important parallels between present-day discourse and practice and what has gone before. Providing this *longue durée* view helps to denaturalize the contemporary crisis, enabling us to see it in a new light.

In chapter 1, Michael R. Dove places contemporary climate change discourse within a context of millennia of human experience with and thinking about climate and society. Examining four canonical historic texts—works by Hippocrates, Theophrastus, Montesquieu, Ibn Khaldûn, and an exploration of Vedic teachings by Zimmermann—Dove traces the continuities and discontinuities in ideas about climate and society. He highlights a number of common concerns that run through these texts: interest in the epistemology of climate knowledge; a focus on human difference versus similarity; an emphasis on the impact of nature on culture, as opposed to the reverse; the pedagogy of peopled versus unpeopled landscapes; and the distancing of climate problems from ourselves. Through this analysis Dove demonstrates the value of historical studies in decentering and thereby illuminating contemporary thinking about climate and society.

The second chapter, by Ben Orlove, Heather Lazrus, Grete Hovelsrud, and Alessandra Giannini, adds further historical depth to our understanding of climate change. The chapter focuses on four regions: the Arctic, islands, deserts, and mountains. While the plight of the Arctic and small island states has come to epitomize the danger of climate change, deserts and mountains are much less commonly associated with the effects of climate change. Orlove and his colleagues explore why this is the case by looking at the historical associations between these places and environmental concerns. In tracing these historical associations, the chapter reveals how discourses of climate change impacts are not the product of powerful nations alone but also reflect the engagement of weak, often disenfranchised speakers from regions distant from centers of power.

Turning to the question of climate change mitigation, chapter 3, by Pamela McElwee, shows how new climate change-inspired land-use policies must be seen in relation to past policies that governed people's interactions with the land. The chapter examines international efforts to mitigate climate change by reducing carbon emissions from deforestation and degradation (REDD+). Drawing on her ethnographic work in Vietnam, McElwee identifies similarities between current programs for REDD+ and previous, usually unsuccessful approaches to forest management. McElwee posits that major barriers to implementation of REDD+ lie in knowledge production about forest environments and forest peoples, and in overly simplistic "checklist" approaches to safeguards and participation—issues that plagued earlier tropical forest management programs. The success of climate change mitigation projects like REDD+, the chapter demonstrates, rests on an appreciation of the fact that these initiatives are not merely about climate governance but also part of a history of struggles over local forest rights.

Part II. Knowing Climate Change

The second part of the book examines the ways in which we "know" climate change. It explores how the body of knowledge about climate change and its impacts is produced and interpreted by scientists, modelers, policy makers, and community members. In these chapters we see how climate change creates both new knowledge and, at the same time, new types of ignorance—something that is often less studied (Mathews 2014; Proctor and Schiebinger 2008). These studies raise the intriguing possibility that knowledge and ignorance of climate change are coproduced—in other words, that the processes of producing knowledge and its opposite in this case are not independent.

In chapter 4, Jessica O'Reilly looks at the Intergovernmental Panel on Climate Change (IPCC), whose reports offer a benchmark of the status of scientific knowledge on climate change. O'Reilly takes us into the details of the IPCC report-writing process, the back and forth of drafting and review and the input of multiple parties. Her chapter examines two controversies surrounding the IPCC Fourth Assessment Report: one arising from a typographical error in the Himalayan glacier melt projection, the other from the exclusion of dynamical processes from the projections of sea level rise. O'Reilly writes the history of how these numbers came to be, tracing the chain of citations and review processes that each reference traveled along. The chapter uncovers how the IPCC assessment report, though peer reviewed, occasionally contains "casual numbers." While not undermining the basic message of the IPCC report, this analysis raises a critique of the phenomenon of assessment itself, showing how it moves from a scientific project to one more akin to auditing and accountability.

Chapter 5, by Jessica Barnes, explores the various forms of knowledge that differently positioned actors draw on in shaping their understandings of resource futures under climate change. Her chapter takes the case of climate change impacts in Egypt, focusing on potential changes in Nile River flows. Barnes examines three contrasting ways of looking at Egypt's future water supply from the Nile: through climate and hydrological modeling, in terms of upstream dam-building projects, and as a function of more localized decisions about water allocation and the operation of engineering structures. These contrasting visions demonstrate how different people attach varying degrees of weight to climate change as an explanatory variable for how much water Egypt will receive from the Nile in the future. The chapter thereby reveals the linkages between the scale at which a future is imagined, the degree to which that future is seen in terms of climate change, and the notions of natural and human agency that underpin those understandings.

In chapter 6, Karina Yager also looks at the question of climate change impacts, contrasting two distinct ways of understanding those impacts. Her chapter focuses on changes in the peatlands of Sajama National Park in Bolivia and compares a satellite imagery analysis of landcover change driven by deglaciation with local Andean herder communities' perceptions of change. The comparison of different knowledge sets on climate change impacts shows both complementary and divergent assessments of change. Yager's analysis indicates, further, how social trends and local management practices also influence peatlands, independently of deglaciation, highlighting the difficulty of isolating climate change as a driving force behind changes in this important landcover type.

In the last chapter in this section, chapter 7, Frances Moore, Justin Mankin, and Austin Becker reflect on the challenges of integrating the diverse forms of knowledge about climate change produced by natural and social scientists. Moore and her colleagues argue that the social sciences are essential for interpreting predictions regarding the physical impacts of climate change in ways that are meaningful to those who have to make decisions in the face of changing climates. Yet statistically based climate projections from the natural sciences are not easily combined with embedded narratives from the social sciences. The chapter explores three critical challenges to integration between the natural and social sciences: the long timescales and large spatial scales at which climate science typically operates; the tendency to abstract impacts related to climate change from the context in which they are manifest; and a strong emphasis on prediction through the use of quantitative models. The authors conclude by outlining a methodological approach which combines the work of anthropologists and other social scientists with that of natural scientists in order to develop more complete and nuanced understandings of climate change and its impacts.

Part III. Imagining Climate Change

The third section of the book draws attention to the role of the imagination in shaping conceptions of climate change. In addition to the material dimension of society–environment relations there is an ideational dimension. The way people think about the environment and about climate change, whether they are government officials, scientists, or peasants, can profoundly influence how they behave. The factors that affect this thinking—politics, education, self-interest, the daily exigencies of life—and the ways in which this thinking is constrained, by culture or memory, for example—are critical to understanding the challenges of contemporary climate change.

In chapter 8, Andrew Mathews looks at how climate change is and has been imagined, through an analysis of forest–climate dynamics in Mexico. Mathews compares two ways of connecting Mexican forests to climate: first, management and planning regimes of industrial forestry beginning in the 1920s; and second, present-day scenario-building practices associated with initiatives to reduce carbon emissions through REDD. He underscores the forms of knowledge and action, and the framings of state, forest, and climate that underlie these two interpretations of forest–climate linkages. The chapter shows how the role of the state has shifted from overseeing calculations of forest cover to convening scenarios about forest futures and insuring deforestation risk across the landscape.

Chapter 9, by Myanna Lahsen, looks at the subcultural tensions that structure climate science politics in the United States. She analyzes both contrarian and more moderate, mainstream critics of anthropogenic climate change science. Lahsen argues that whatever the truth value of their criticisms, competition for status and research funds are an important subtext, as are experiences of marginalization and alienation stemming from the rise of environmental concerns in the 1980s and a related emphasis on environmental sciences and simulation techniques. Lahsen concludes that in order to identify the factors that shape scientists' perceptions of climate change and the nature of their engagements with public debates about climate science, it is necessary to probe beyond individual characteristics (such as those emphasized in public opinion surveys on climate change) to look at these kinds of embedded, cultural dynamics.

In chapter 10, Rajindra Puri focuses on everyday actions and interactions between people and their environments. The realm of the everyday is the one in which impacts of climate change will be experienced and adaptive responses will become apparent. Puri looks at Indian herders' response to an invasive species, *Lantana camara*, whose proliferation may be linked to changing precipitation patterns. He uses this to open a window into human decision making in the face of environmental change. His chapter illustrates how a focus on quotidian activities can inform understandings of adaptation to climate change impacts.

Chapter 11, by Roderick McIntosh, explores the idea of deep-time cultural memory of climatic variability and responses to it. His argument centers on the Mande of arid West Africa, who live with precipitation and fluvial regimes that are among the least predictable in the world. Fluctuation in these regimes occurs over such long timescales as to be impossible to retain within conscious cultural memory. McIntosh asks how ancient peoples understood climatic changes, how those understandings shaped their actions, and how they passed the experience thereby acquired on to distant future generations. Contemporary Mande's beliefs and values about these deep-time lessons of the world in which their ancestors lived, McIntosh argues, constitute the core of an ethno-science of climate. In featuring the role of Mande "weather machines"—people with the moral authority to persuade others—the chapter reflects on the question of who is authorized, in either the current or the prehistoric moment, to make decisions about how to act in the face of climate change, and how authority and knowledge are conveyed over intergenerational time periods.

In the afterword, Mike Hulme draws together insights from all of the chapters around themes of causation, representation, and instrumentalism. He explores the role of climate as an index, as embodied in the term *climatic change*, versus climate as an agent, as in the term *climate change*. In the former, climate is a descriptor of change; in the latter, it is a cause of change. This distinction signifies, Hulme argues, an underlying debate over the balance between human and natural agency.

Taken as a whole, the contents of this book illustrate what we regard as the unique contribution that anthropology can make to the contemporary study of climate change. Our purpose in writing the book is to broaden the intellectual effort to address one of the most pressing problems of our time. In many popular and political debates, climate change appears to be a nearly intractable problem. There may indeed be dimensions of it that will elude our solutions. But anthropological research can make at least parts of the problem more tractable, can help us understand some of the more incomprehensible issues, and can help explain why other parts remain so intractable.

NOTES

1. Cover of the *New Yorker* magazine for August 13 and 20, 2012.
2. Our use of the term *holistic* to describe a positive attribute of anthropology as a discipline departs from Hastrup (2013), who uses the term to refer to a dated anthropology, in which culture is seen as bounded, isolated, and static.

REFERENCES CITED

Agrawal, A., M. Lemos, B. Orlove, and J. Ribot. 2012. "Cool Heads for Hot World—Social Sciences under a Changing Sky." *Global Environmental Change* 22: 329–31.

Anderson, D., K. Maasch, and D. Sandweiss, eds. 2006. *Climate Change and Cultural Dynamics: A Global Perspective on Mid-Holocene Transitions.* New York: Elsevier.

Axtell, R., J. Epstein, J. Dean, G. J. Gumerman, A. C. Swedlund, J. Harburger, S. Chakravarty, R. Hammond, J. Parker, and M. Parker. 2002. "Population Growth and Collapse in a Multiagent Model of the Kayenta Anasazi in Long House Valley." *Proceedings of the National Academy of Science* 99: 7275–79.

Bar-Yosef, O. 2011. "Climatic Fluctuations and Early Farming in West and East Asia." *Current Anthropology* 52: S175–S193.

Barnes, J., M. Dove, M. Lahsen, A. Mathews, P. McElwee, R. McIntosh, F. Moore, J. O'Reilly, B. Orlove, R. Puri, H. Weiss, and K. Yager. 2013. "Contribution of Anthropology to the Study of Climate Change." *Nature Climate Change* 3: 541–44.

Batterbury, S. 2008. "Anthropology and Global Warming: The Need for Environmental Engagement." *Australian Journal of Anthropology* 19: 62–68.

Bee, B. 2013. "Who Reaps What Is Sown? A Feminist Inquiry into Climate Change Adaptation in Two Mexican Ejidos." *ACME: An International E-Journal for Critical Geographies* 12: 131–54.

Bettini, G. 2013. "Climate Barbarians at the Gate? A Critique of Apocalyptic Narratives on 'Climate Refugees.'" *Geoforum* 45: 63–72.

Bjurström, A., and M. Polk. 2011. "Physical and Economic Bias in Climate Change Research: A Scientometric Study of the IPCC Third Assessment Report." *Climatic Change* 108: 1–22.

Brondizio, E., and E. Moran. 2008. "Human Dimensions of Climate Change: The Vulnerability of Small Farmers in the Amazon." *Philosophical Transactions of the Royal Society of London B* 363: 1803–9.

Bunker, S. 1985. *Underdeveloping the Amazon: Extraction, Unequal Exchange, and the Failure of the Modern State.* Urbana: University of Illinois Press.

Callon, M., J. Law, and A. Rip. 1986. *Mapping the Dynamics of Science and Technology: Sociology of Science in the Real World.* Basingstoke: Macmillan.

Cassidy, R. 2012. "Lives with Others: Climate Change and Human–Animal Relations." *Annual Review of Anthropology* 41: 21–36.

Chakrabarty, D. 2009. "The Climate of History: Four Theses." *Critical Inquiry* 35: 197–222.

Cockcroft, J., A. Frank, and D. Johnson. 1972. *Dependence and Underdevelopment.* New York: Anchor Books.

Collins, H., and T. Pinch. 1982. *Frames of Meaning: The Social Construction of Extraordinary Science.* London: Routledge.

Crate, S. 2008. "Gone the Bull of Winter: Grappling with the Cultural Implications of and Anthropology's Role(s) in Global Climate Change." *Current Anthropology* 49, no. 4: 359–95.

———. 2011. "Climate and Culture: Anthropology in the Era of Contemporary Climate Change." *Annual Review of Anthropology* 40: 175–94.

Crate, S., and M. Nuttall. 2009. *Anthropology and Climate Change: From Encounters to Action.* Walnut Creek, Calif.: Left Coast Press.

Davidson, J. 2013. "Of Rice and Men: Climate Change, Religion, and Personhood among the Diola of Guinea-Bissau." *Journal for the Study of Religion, Nature, and Culture* 6, no. 3: 363–81.

Diemberger, H., K. Hastrup, S. Schaffer, C. F. Kennel, D. Sneath, M. Bravo, H. F. Graf, J. Hobbs, J. Davis, M. L. Nodari, G. Vassena, R. Irvine, C. Evans, M. Strathern, M. Hulme, G. Kaser, and B. Bodenhorn. 2012. "Communicating Climate Knowledge: Proxies, Processes, Politics." *Current Anthropology* 53: 226–44.

Diffenbaugh, N., and M. Scherer. 2013. "Likelihood of July 2012 U.S. Temperatures in Pre-Industrial and Current Forcing Regimes." *Bulletin of the American Meteorological Society* 94: S6–S9.

Dillehay, T., and A. Kolata. 2004. "Long-Term Human Response to Uncertain Environmental Conditions in the Andes." *Proceedings of the National Academy of Science* 101: 4325–30.

Dove, M. 1994. "North–South Relations, Global Warming, and the Global System." *Chemosphere* 29: 1063–77.

Farbotko, C., and H. Lazrus. 2012. "The First Climate Refugees? Contesting Global Narratives of Climate Change in Tuvalu." *Global Environmental Change* 22, no. 2: 382–90.

Ferguson, J. 2006. *Global Shadows: Africa in the Neoliberal World Order*. Durham: Duke University Press.

Fleming, J. 1998. *Historical Perspectives on Climate Change*. New York: Oxford University Press.

Fleming, J., and V. Jankovic. 2011. "Revisiting Klima." *Osiris* 26: 1–15.

Frank, A. 1967. *Capitalism and Underdevelopment in Latin America: Historical Studies of Chile and Brazil*. New York: Monthly Review Press.

Geertz, C. 1973. *The Interpretation of Cultures: Selected Essays*. New York: Basic Books.

Günel, G. 2012. "A Dark Art: Field Notes on Carbon Capture and Storage Policy Negotiations at COP17." *Ephemera: Notes in Organization Theory and Politics* 12: 33–41.

Gupta, A., and J. Ferguson. 1997. "Discipline and Practice: 'The Field' as Site, Method, and Location in Anthropology." In *Anthropological Locations: Boundaries and Grounds of a Field Science*, ed. A. Gupta and J. Ferguson, 1–46. Berkeley: University of California Press.

Hamblyn, R. 2009. "The Whistleblower and the Canary: Rhetorical Constructions of Climate Change." *Journal of Historical Geography* 35: 223–36.

Hastrup, K. 2009. "Arctic Hunters, Climate Variability and Social Flexibility." In *The Question of Resilience: Social Responses to Climate Change*, ed. K. Hastrup, 245–70. Copenhagen: Royal Danish Academy of Letters and Sciences.

———. 2012. "Scales of Attention in Fieldwork: Global Connections and Local Concerns in the Arctic." *Ethnography* 14: 145–64.

———. 2013. "Anthropological Contributions to the Study of Climate: Past, Present, Future." *WIREs Climate Change* 4: 269–81.

Hastrup, K., and M. Skrydstrup, eds. 2013. *The Social Life of Climate Change Models: Anticipating Nature*. New York: Routledge.

Henshaw, A. 2009. "Sea Ice: The Sociocultural Dimensions of a Melting Environment in the Arctic." In *Anthropology and Climate Change: From Encounters to Actions*, ed. S. Crate and M. Nuttall, 153–65. Walnut Creek, Calif.: Left Coast Press.

Hulme, M. 2011a. "Meet the Humanities." *Nature Climate Change* 1: 177–79.

———. 2011b. "Reducing the Future to Climate: A Story of Climate Determinism and Reductionism." *Osiris* 26: 245–66.

IPCC. 2013. *Climate Change 2013: The Physical Science Basis: Working Group I Contribution to the Fifth Assessment Report of the Intergovernmental Panel on Climate Change (IPCC)*, ed. T. Stocker, D. Qin, G.-K. Plattner, M. Tignor, S. K. Allen, J. Boschung, A. Nauels, Y. Xia, V. Bex and P. M. Midgley. New York: Cambridge University Press.

Jasanoff, S. 2004. *States of Knowledge: The Co-Production of Science and the Social Order*. London: Routledge.

———. 2010. "A New Climate for Society." *Theory, Culture & Society* 27, no. 2–3: 233–253.

Jonsdottir, A. 2013. "Scaling Climate: The Politics of Anticipation." In *The Social Life of Climate Change Models: Anticipating Nature*, ed. K. Hastrup and M. Skrydstrup, 128–43. New York: Routledge.

Kaniewski, D., E. van Campo, and H. Weiss. 2012. "Drought Is a Recurrent Challenge in the Middle East." *Proceedings of the National Academy of Science* 109: 3862–67.

Knorr Cetina, K. 1999. *Epistemic Cultures: How the Sciences Make Knowledge.* Cambridge: Harvard University Press.

Lahsen, M. 2002. "Brazilian Climate Epistemers' Multiple Epistemes: An Exploration of Shared Meaning, Diverse Identities, and Geopolitics in Global Change Science." Belfer Center for Science and International Affairs (BCSIA) Discussion Paper 2002–01. Cambridge, Mass.: Environment and Natural Resources Program, Kennedy School of Government, Harvard University.

———. 2005. "Seductive Simulations: Uncertainty Distribution Around Climate Models." *Social Studies of Science* 35: 895–922.

———. 2007. "Anthropology and the Trouble of Risk Society." *Anthropology News* (December): 9–10.

———. 2009. "A Science–Policy Interface in the Global South: The Politics of Carbon Sinks and Science in Brazil." *Climatic Change* 97: 339–72.

Lansing, D. 2011. "Realizing Carbon's Value: Discourse and Calculation in the Production of Carbon Forestry Offsets in Costa Rica." *Antipode* 43: 731–53.

Latour, B., and S. Woolgar. 1986. *Laboratory Life.* Princeton: Princeton University Press.

Lipset, D. 2011. "The Tides: Masculinity and Climate Change in Coastal Papua New Guinea." *Journal of the Royal Anthropological Institute* 17: 20–43.

Magistro, J., and C. Roncoli. 2001. "Anthropological Perspectives and Policy Implications of Climage Change Research." *Climate Research* 19: 91–96.

Mahony, M. 2014. "The Predictive State: Science, Territory and the Future of the Indian Climate." *Social Studies of Science* 44, no. 1: 109–33.

Marcus, G. 1995. "Ethnography in/of the World System: The Emergence of Multisited Ethnography." *Annual Review of Anthropology* 24: 95–117.

Marino, E., and P. Schweitzer. 2009. "Talking and Not Talking about Climate Change in Northwestern Alaska." In *Anthropology and Climate Change: From Encounters to Actions,* ed. S. Crate and M. Nuttall, 209–17. Walnut Creek, Calif.: Left Coast Press.

Mathews, A. 2009. "Unlikely Alliances: Encounters Between State Science, Nature Spirits, and Indigenous Industrial Forestry in Mexico, 1926–2008." *Current Anthropology* 50: 75–101.

———. 2014. "Scandals, Audits, and Fictions: Linking Climate Change to Mexican Forests." *Social Studies of Science* 44, no. 1: 82–108.

McElwee, P. 2010. "Resource Use Among Rural Agricultural Households Near Protected Areas in Vietnam: The Social Costs of Conservation and Implications for Enforcement." *Environmental Management* 45: 113–31.

McIntosh, R., J. Tainter, and S. McIntosh. 2000. *The Way the Wind Blows: Climate, History and Human Action.* New York: Columbia University Press.

Moore, A. 2010. "Climate Change and Small Islands: Considering Social Science and the Production of Island Vulnerability and Opportunity." *Environment and Society: Advances in Research* 1: 116–31.

Moore, F. 2012. "Negotiating Adaptation: Norm Selection and Hybridization in International Climate Negotiations." *Global Environmental Politics* 12, no. 4: 30–48.

Nuttall, M. 2009. "Living in a World of Movement: Human Resilience to Environmental Instability in Greenland." In *Anthropology and Climate Change: From Encounters to Actions*, ed. S. Crate and M. Nuttall, 292–310. Walnut Creek, Calif.: Left Coast Press.

O'Reilly, J., N. Oreskes, and M. Oppenheimer. 2012. "The Rapid Disintegration of Projections: The West Antarctic Ice Sheet and the Intergovernmental Panel on Climate Change." *Social Studies of Science* 42: 709–31.

Orlove, B. 2005. "Human Adaptation to Climate Change: A Review of Three Historical Cases and Some General Perspectives." *Environmental Science and Policy* 8: 589–600.

———. 2009. "The Past, the Present and Some Possible Futures of Adaptation." In *Adapting to Climate Change: Thresholds, Values, Governance*, ed. N. Adger, I. Lorenzoni, and K. O'Brien, 131–63. Cambridge: Cambridge University Press.

Orlove, B., J. Chiang, and M. Cane. 2000. "Forecasting Andean Rainfall and Crop Yield from the Influence of El Niño on Pleiades Visibility." *Nature* 403: 68–71.

———. 2002. "Ethnoclimatology in the Andes." *American Scientist* 90: 428–35.

Parker, G. 2013. *Global Crisis: War, Climate Change, and Catastrophe in the Seventeenth Century*. New Haven: Yale University Press.

Pickering, A. 1995. *The Mangle of Practice: Time, Agency, and Science*. Chicago: University of Chicago Press.

Proctor, R., and L. Schiebinger, eds. 2008. *Agnotology: The Making and Unmaking of Ignorance*. Stanford: Stanford University Press.

Rosen, A. 2007. *Civilizing Climate: Social Responses to Climate Change in the Ancient Near East*. Lanham, Md.: AltaMira Press.

Rudiak-Gould, P. 2012. "Promiscuous Corroboration and Climate Change Translation: A Case Study from the Marshall Islands." *Global Environmental Change* 22: 46–54.

———. 2013. *Climate Change and Tradition in a Small Island State: The Rising Tide*. New York: Routledge.

Sahlins, M. 2004. *Apologies to Thucydides: Understanding History as Culture and Vice Versa*. Chicago: University of Chicago Press.

Sorlin, S. 2009. "Narratives and Counter-Narratives of Climate Change: North Atlantic Glaciology and Meteorology, c. 1930–1955." *Journal of Historical Geography* 35, no. 2: 237–55.

Strauss, S., and B. Orlove. 2003. "Up in the Air: The Anthropology of Weather and Climate." In *Weather, Climate, Culture*, ed. S. Strauss and B. Orlove, 3–14. London: Berg.

Thornton, T., and N. Manasfi. 2010. "Adaptation—Genuine and Spurious: Demystifying Adaptation Processes in Relation to Climate Change." *Environment and Society: Advances in Research* 1: 132–55.

Townsend, P. 2010. "Still Fiddling while the Globe Warms?" *Reviews in Anthropology* 33: 335–49.

Tsing, A. L. 2005. *Friction: An Ethnography of Global Connection*. Princeton: Princeton University Press.

Vinthagen, S. 2013. "Ten Theses on Why We Need a Social Science Panel on Climate Change." *ACME: An International E-Journal for Critical Geographies* 12: 155–76.

Weiss, H., and R. Bradley. 2001. "What Drives Societal Collapse?" *Science* 291, no. 5504: 609–10.

Weiss, H., A. Courty, W. Wetterstrom, F. Guichard, L. Senior, R. Meadow, and A. Curnow. 1993. "The Genesis and Collapse of Third Millennium North Mesopotamian Civilization." *Science* 261: 995–1004.

Wolf, E. 1982. *Europe and the People Without History.* Berkeley: University of California Press.

Yearley, S. 2009. "Sociology and Climate Change after Kyoto: What Role of Social Science in Understanding Climate Change?" *Current Sociology* 57: 389–405.

Part One **Historicizing Climate Change**

Chapter 1 **Historic Decentering of the Modern Discourse of Climate Change: The Long View from the Vedic Sages to Montesquieu**

Michael R. Dove

We have always thought about climate. However unique modern climate change may be, a discourse of climate and culture has been prominent within human society for millennia. Indeed, it might be said to have been an integral part of *the* discourse of civilization itself (Glacken 1967: vii). My thesis in this chapter is that the millennia of human experience with the climate and the millennia of human thinking about the climate are relevant to the contemporary crisis of climate change. There has been considerable work on the historic experience of climate by archaeologists and prehistorians (e.g., Weiss 2000); there has been little if any work on the premodern history of climate-related thinking (McIntosh 2000 is an exception; see also McIntosh, this volume).

I will first discuss the conflicted stance vis-à-vis history that characterizes the contemporary discourse of climate change and the associated proposition that climate change heralds the "end of history." Then I will briefly present the intellectual contributions to thinking about climate and society of five significant bodies of work: the Hippocratic tradition; its revival in the work of Montesquieu; the related work of the medieval Islamic historian Ibn Khaldûn; the work of the Vedic sages as interpreted by Zimmermann; and the work of Theophrastus, a near-contemporary of Hippocrates. Next I will analyze continuities and discontinuities in thinking regarding climate and society over the past several millennia, focusing on epistemology, ideas about nature and culture, peopled versus unpeopled landscapes, the emphasis on difference versus similarity, and ideas of the other. I will conclude with a discussion of what we can achieve through such historical comparisons, and what this reveals about the potential contribution of anthropology to contemporary climate change scholarship.

I am not attempting here a history of thinking about climate and society. I selected the aforementioned historic scholars because they are exemplary, illuminating, and influential in that thinking for centuries, if not millennia. I begin with Hippocrates because of his historic influence on the Western tradition. I include Theophrastus as a contemporary of Hippocrates who stands out for his study of indigenous knowledge of climate. I include Zimmermann and the Vedic texts because they are both similar to and different from the scholarship of the Greeks. I present Ibn Khaldûn as an application of Hippocratic thought within the Islamic academic tradition. Finally, I cite Montesquieu as an application of Hippocratic thought to the novel realm of law and politics, one that is so comparatively modern as to testify to the long-lasting impact of Hippocrates and the relative recency of the scientific paradigm that replaced it. These studies represent, therefore, multiple intellectual traditions: East and West; Christian, Hindu, and Muslim; modern and premodern.

AMBIVALENCE REGARDING HISTORY

Contemporary discussion of climate change is thoroughly imbricated with a sense of history. Participants look not only to the future and its threat of catastrophe but also to the past and its perceived lessons. Climate scientists look at the long history of the earth's climate for baselines and analogs to inform their model building; archaeologists look at the dual histories of climate and society over the past millennia to speculate on interrelationships; and historians look at the El Niño–Southern Oscillation (ENSO) phenomenon and perturbations like the so-called Little Ice Age to understand social change over the past centuries. There is a schizophrenic twist to the role played by history in the climate change discourse, however: whereas climate scientists draw on historical data to inform their research, so too do climate change deniers. The historic fluctuations in climate that climate scientists cite in support of their work are invoked by deniers as evidence that contemporary changes in climate fall within historic norms. Whereas the climate science community is using history to exceptionalize the present, the deniers are using history to normalize it. Indeed, because of the arguments being made by the denial camp, some scientists working on climate change now regard the historic perspective as an inherently apologist one. This is making some climate scientists skittish with regard to historical studies, as indeed some have been from the beginning (Demeritt 2001: 315). Some scholars of climate change are reasoning that if they cannot build a case for the exceptionalness of modern climate change based on history, if history is not going to help, then history should be ruled out of the equation. It should be ended. But this raises the question, as Derrida (1994) writes: "How can one be late to the end of history? . . . [I]t obliges one to wonder if the end of history is but the end of a certain concept of history. Here is perhaps one of the questions that should be asked of those who are not content just to arrive late to the apocalypse and to the last train of the end. . . . [T]here where history is finished, there where a certain determined concept of history comes to an end, precisely there the historicity of history begins."

This stance on history is reflected most dramatically in the coining of the term *Anthropocene* (succeeding the Holocene) a decade ago in recognition of the new role of humans as a "global forcing agent" (Crutzen 2002). Although this term and the idea of a historic break in geologic time that it signifies has perhaps found more favor with social scientists than with natural scientists (but see Zalasiewicz et al. 2010), it has gained wide currency in the field. Chakrabarty (2009) sums up some of the implications of this break from the perspective of a historian. Whereas some interpret the global scale of unintended "geo-engineering" of the climate as amplifying the nature/culture dichotomy (cf. Ingold 1993), Chakrabarty (2009: 201) sees it as causing "the collapse of the age-old humanist distinction between natural history and human history." Whereas the traditional idea of history was based on the idea of social humans living separate from nature, the climate-changed future presents the idea of geological humans or humans as species, who are one with nature (Chakrabarty 2009: 203, 205; cf. Crutzen 2002; Wilson 1996).[1] As Chakrabarty writes, "In unwittingly destroying the artificial but time-honored distinction between natural and human histories, climate scientists posit that the human being has become something much larger than the simple biological agent that he or she always has been" (2009: 206). Chakrabarty sees this development as historically unique: self-understanding as a species "does not correspond to any historical way of understanding and connecting pasts with futures through the assumption of there being an element of continuity to human experience" (2009: 220). Of most importance to Chakrabarty, as a historian, this puts the future "beyond the grasp of historical sensibility" (2009: 197).

CANONICAL HISTORIC WORKS ON CLIMATE AND SOCIETY

The concept of the Anthropocene and the clean historic break clearly has value in helping to mobilize thinking and resources in the battle against the causes and consequences of climate change. But it bears costs as well, and one that has not heretofore been noted is that of separating us from the intellectual tradition of thinking about climate and society. The engagement by the climate science community with history to date has been with the history of climate, not with the history of ideas about climate; and the concept of the Anthropocene will not remedy this. If anthropogenic climate change is the end of history, as Chakrabarty says, then what went before is no longer relevant. But is this true? What would happen if current thinking about climate was informed by past thinking about climate (see McIntosh, this volume)? This is a question worth asking, not least because the contemporary denialist stance, although dismissed by the climate science community, has been gaining, not losing, traction in U.S. politics in recent years.

Hippocrates, "Airs, Waters, Places"

There are important commentaries on environmental matters in the writings of many scholars of the classical era, including Aristotle, Herodotus, Thucydides, and Pliny. But in terms of an extended, in-depth analysis of the relationship between culture and nature, the work of Hippocrates, born in 460 BC (Jones 1923: xliii), is perhaps unsurpassed. His "Airs, Waters, Places" is a seminal work on the linkages between climate, landscape, physique, and temperament and has been called "the earliest systematic treatise concerned with environmental influences on human culture" (Glacken 1967: 81–82). Hippocrates (1923: 109) writes, "For where the seasons experience the most violent and the most frequent changes, the land too is very wild and uneven; you will find there many wooded mountains, plains, and meadows. But where the seasons do not alter much, the land is very even. So it is too with the inhabitants, if you will examine the matter. Some physiques resemble wooded, well-watered mountains, others light, dry land, others marshy meadows, others a plain of bare, parched earth."

"Airs, Waters, Places" has separate chapters for biological and cultural human beings, that is, for medicine and ethnography. In the first, medical part of this essay, Hippocrates (1923: 105) relates human health to the characteristics of the site or locale and its seasons; in the second, ethnographic part, he generalizes from this causal relationship to explain the correlation between the character of entire societies and the regions within which they live: "Now I intend to compare Asia and Europe, and to show how they differ in every respect, and how the nations of the one differ entirely in physique from those of the other." The Hippocratic tradition presents, therefore, two distinct but related bodies of theory regarding disease and locale on the one hand, and on the other hand society and environment. The differences and similarities between the two parts can be illustrated in terms of a series of paired correspondences:

body: society :: medicine: ethnography :: seasons: geography

The independent variable in the case of physiology is locale- and seasonally driven variation in weather; the independent variable in the case of geography is latitudinally and topographically driven variation in climate. The intent of "Airs, Waters, Places" is prognostic: the aim in the first, medical part is to predict the effects of time and place on human health; the intent of the second, ethnographic part is to predict the effect of climatic regions on human character.

Charles de Secondat Montesquieu, *The Spirit of the Laws* (*Esprit des loix*)

The Hippocratic work on climate and society was revived during the Enlightenment by Montesquieu in his *The Spirit of the Laws*, the writing in which is clearly reminiscent of "Airs, Waters, Places" (1752: 190): "If we travel towards the north, we meet with people who have few vices, many virtues, and a great share of frankness and sincerity. If we draw near the south, we fancy ourselves entirely removed from the verge of morality; here the strongest passions are productive of all manner of crimes, each man endeavoring, let the means be what they will, to indulge his inordinate desires." Montesquieu, who read widely (Cohler, Miller, and Stone 1989: xx), does not specifically refer to Hippocrates in *The Spirit of the Laws*, but he is known to have had a version of "Airs, Waters, Places" in his library, and its impact on his work is generally acknowledged (Levin 1936: 26–39).

Montesquieu was born in 1689 at the Château de La Brède in southwestern France and died in 1755 in Paris. He was variously a "feudal proprietor, wine merchant, parliamentarian, academician, man of letters" (Cohler, Miller, and Stone 1989: xii). He published *The Spirit of the Laws* in 1748 and considered it to be his lifework. As the first "systematic treatise on politics" since Aristotle, it is justly famed for its principle of separation of powers for securing liberty and for the impact this had on constitutionalism in the United States and elsewhere around the world (Neuman 1949: x, lix).

Montesquieu is seen by many anthropologists as an intellectual forebear (Launay 2010). For example, he displays in his writings a marked element of cultural relativism with respect to non-Christian religions and practices like polygamy (Nugent 1752: 6–7), which, among other aspects of *The Spirit of the Laws*, got him in political trouble in France and obliged him to publish a subsequent defense of it (Neuman 1949: xiii). Equally relevant to anthropologists and of most importance here is Montesquieu's contribution to theorizing about the relationship between nature and culture, in particular between climate and law: "If it be true that the temper of the mind and the passions of the heart are extremely different in different climates, the laws ought to be in relation both to the variety of those passions and to the variety of those tempers" (1752: 188). His thesis that lawmaking should be suited to the character of the society, and that this is in turn influenced by the character of the environment and its climate, can also be traced to Aristotle and Plato. Glacken (1967: 653) asserts, "By his advocacy of climatic influences, Montesquieu in the *Esprit des Lois* had provoked some of the most searching thought on social and environmental questions that had yet appeared in Western civilization."

Ibn Khaldûn, *The Muqaddimah: An Introduction to History*

From yet another, not entirely unrelated tradition derives the work by the medieval Islamic scholar Ibn Khaldûn, *The Muqaddimah*, which has been called "the most significant and challenging Islamic history of the premodern world" (Lawrence 2005: vii). Born in 1332 in Andalusia in southern Spain in a Muslim family that had migrated there from Yemen in the eighth century, Ibn Khaldûn died in 1406 in Cairo. Formally trained as a *faqîh*, or jurist, who spent his life in public, political service, Ibn Khaldûn was also an *adîb*, or man of letters. He is still known today for his seminal work, *The Muqaddimah: An Introduction to History*, published in Arabic in 1370 in Cairo and described by Arnold Toynbee (1935: 3:322) as "undoubtedly the greatest work of its kind that has ever yet been created by any mind in any time or place." Written at the end of the intellectual development of medieval Islam, it captured the historical depth and conceptual heights of this development (Rosenthal 1958: cxiii).

The medieval Islamic renaissance was distinct from but not unconnected to Western intellectual traditions, and the links to classical Greek scholarship were explicit. As Lawrence (2005: xi) writes of Ibn Khaldûn, "He engaged the full spectrum of sciences that were known in Arabic translations from Greek sources by the ninth century." Indeed, Ibn Khaldûn writes that "the sciences of only one nation, the Greeks, have come down to us, because they were translated through al-Ma'mûn's efforts" (1958: 1:78), an 'Abbâsid caliph reputed to have sent emissaries to the Byzantine emperors to find and copy Greek manuscripts (3:116). He refers in multiple places to translations of Greek works (1958: 2:203, 3:130, 151, 250); and he explicitly states that "Muslim scientists assiduously studied the (Greek sciences)" (1958: 3:116). The resemblance to the Hippocratic writings is clear:

> The inhabitants of the zones that are far from temperate, such as the first, second, sixth, and seventh zones [those closest to either the equator or the North Pole], are also farther removed from being temperate in all their conditions. Their buildings are of clay and reeds. Their foodstuffs are dura [grain] and herbs. Their clothing is the leaves of trees, which they sew together to cover themselves, or animal skins. Most of them go naked. The fruits and seasonings of their countries are strange and inclined to be intemperate. . . . Their qualities of character, moreover, are close to those of dumb animals. It has even been reported that most of the Negroes of the first zone dwell in caves and thickets, eat herbs, live in savage isolation and do not congregate, and eat each other. The same applies to the Slavs. The reason for this is that their remoteness from being temperate produces in them a disposition and character similar to that of the dumb animals, and they become correspondingly remote from humanity. The same also applies to their religious conditions. They are ignorant of prophecy and do not have a religious law, except for the small minority that lives near the temperate regions. (Ibn Khaldûn 1958: 1:168)

Ibn Khaldûn is another of the early figures claimed as ancestors of modern anthropology, based on his theorizing regarding the dynastic cycles of the Islamic states of North Africa, which he claimed characteristically run their full course—from ascent to decline—in just three generations (Rosenthal

1958: lxxxii; cf. Launay 2010). The driver of this dynastic cycle is the dynamic relationship between the two fundamentally contrasting socio-ecologies of the region—the urban and the hinterland. Going beyond Hippocrates, therefore, Ibn Khaldûn does not just distinguish the two socio-ecologies but examines the relationship between them.

Francis Zimmermann, *The Jungle and the Aroma of Meats: An Ecological Theme in Hindu Medicine*

Largely outside the Greek and Islamic traditions is an older Vedic tradition from the subcontinent, a tradition that has its own unique development of climate theory. The Sanskrit texts on which this tradition is based were composed two thousand to four thousand years ago; but they are living texts, still cited today in Ayurvedic teachings. Francis Olivier Zimmermann, born in France in 1942, is an anthropologist and currently Directeur d'études à l'Ecole des Hautes Etudes en Sciences Sociales in Paris. His book *The Jungle and the Aroma of Meats: An Ecological Theme in Hindu Medicine* (published in 1987 and summarized in Zimmermann 1988) is based on a close reading of the ancient Sanskrit texts and their ecological significance: "Although this book borrows numerous illustrations from tropical geography and contemporary studies of flora and zoology, the reader should not be misled. Its point of departure is not observation but a study of the texts; not geography or the natural sciences but a corpus of traditional notions into which it became possible to subsume the empirical data" (Zimmermann 1987: 3–4).

Zimmermann's work concerns another climate-related dichotomy, like desert versus city and temperate versus intemperate zone. He claims that the ancient Vedic texts describe a cosmological divide between the semi-arid savanna (*jāngala*) of western India and the perennially wet forests (*anūpa*) of eastern India, which is in turn based on a fundamental underlying polarity between *agni* (fire) and *soma* (water). Like the climatic divides discussed earlier, this one has a normative dimension: one zone is the abode of the civilized Aryan, the other is that of the uncivilized barbarians. As Zimmermann writes (1987: 18), "In ancient India, *all the values of civilization lay on the side of the jungle.* The *jāngala* incorporated land that was cultivated, healthy, and open to Aryan colonization, while the barbarians were pushed back into the *anūpa*, the insalubrious, impenetrable lands." The polarity between the savanna and the rain forest "is a matter not of physical facts but of brahminic norms" and thus is not just descriptive but prescriptive (Zimmermann 1987: 39). As stated in *The Laws of Manu*, for example, "Let him [the Hindu king] take up residence in a *jāngala* place, where cereals are abundant, where the ārya predominate and which is free from disorder" (Zimmermann 1987: 39).

Theophrastus, *Enquiry Into Plants and Minor Works on Odours and Weather Signs*

There are many classical texts on the subject of meteorology, including Virgil's (2004) *Georgics*, Pliny's (1942) *Natural History*, and Aristotle's (1952) *Meteorologica*. Specifically having to do with the signs of weather, the major classical works are Hesiod's *Works and Days* (1914) and Theophrastus's (1926) "Concerning Weather Signs." The two differ in that Hesiod focused on the meaning of regular occurrences for the annual weather cycle whereas Theophrastus focused on the meaning of irregular occurrences for immediate weather conditions (Sider and Brunschön 2007: 3–4). As was typical of these classical works, neither one addressed long-term changes in climate. Nor were any of these ancient works self-conscious, ethno-climatological studies, although they all drew and reported on what was essentially local folk knowledge.

Theophrastus was born about 370 BC in Eresos in Lesbos and died around 285 BC. He was a student of Plato and then Aristotle. His *Enquiry Into Plants*, a wide-ranging work of natural history, contains a chapter entitled "Concerning Weather Signs." This is a listing of all of the then-known signs—primarily from either astronomical phenomena or animal behavior—from which weather could be forecast in the short term. The work is structured as a prologue (passages 1–9), followed by signs of rain (10–25), wind (26–37), storms (38–49), and fair weather (50–55). Some of the signs have seasonal implications, but most address weather expected within the next few hours or days. The forecasts are strictly meteorological in character, that is, they predict changes in weather, not, at least not directly, changes in the fortunes of humans (Sider and Brunschön 2007: 36).

"Concerning Weather Signs" is rich in what is today called thick description and presents an empirically flavored exegesis of local knowledge about the weather. Theophrastus writes, "It is a sign of rain or storm when birds which are not aquatic take a bath. It is a sign of rain when a toad takes a bath, and still more so when frogs are vocal. So too is the appearance of the lizard known as 'salamander,' and still more the chirruping of the green frog in a tree. It is a sign of rain when swallows hit the water of the lakes with their belly. It is a sign of storm or rain when the ox licks his fore-hoof; if he puts his head up towards the sky and sniffs the air, it is a sign of rain" (1926:

2:399). This writing is based on a celebration of fine-grained knowledge of the sort that environmental anthropologists and human ecologists have treated as major discoveries in recent decades. Further examples include Theophrastus's observations that "it is a sign of rain if ants in a hollow place carry their eggs up from the ant-hill to the high ground, a sign of fair weather if they carry them down" and "It is a sign of storm . . . when sheep or birds fight for their food more than usual, since they are then trying to secure a store against bad weather" (1926: 2:405, 421). As is the case with what is reported in most good studies in environmental anthropology and human ecology, this knowledge is locally grounded and place-specific. Theophrastus (1926: 2:403, 405, 407) makes multiple references to the observable conditions on the flanks and summits of particular, named mountains. He explicitly underscores the importance of this orientation: "Wherefore good heed must be taken to the local conditions of the region in which one is placed" (Theophrastus 1926: 2:393).

CONTINUITIES: COMPARISONS AND CONTRASTS

What continuities as well as discontinuities exist between the current climate change discourse and these historic texts?

Epistemology

For anthropologists, "thinking about thinking" about climate change is an important subject of study.

CONSTRUCTION OF KNOWLEDGE Central to the contemporary debate about the reality of global climate change are questions about the quality of the evidence, the rigor of the analysis, and the motives of climate scientists (see Lahsen, this volume). Demeritt (2001: 328) suggests that a positive feedback cycle has led climate scientists to overstate their case in response to skepticism regarding climate change, which has only increased the skepticism.

Awareness of the need for evidence and reflection on how to build knowledge are ancient concerns. One translator of Hippocrates's (1923: 66) "Airs, Waters, Places," says, "In tone it is strikingly dogmatic, conclusions being enunciated without the evidence upon which they are based." But throughout the work there is more than a nod toward articulation of a methodology. Hippocrates (1923: 71) begins the book with the explicit direction, "Whoever wishes to pursue properly the science of medicine must proceed thus." He then says that one must consider the effects of the different seasons, the different winds, and the different waters and also different soils and modes of life (1923: 71). As a practical matter, "on arrival at a town with which he is unfamiliar, a physician should examine its position with respect to the winds and the risings of the sun" (1923: 71). Hippocrates continues in this manner, two chapters later writing, "I will now set forth clearly how each of the foregoing questions ought to be investigated, and the tests to be applied" (1923: 73–75). One famous test involves freezing and then thawing water, which purports to show the loss of its "original nature" (1923: 93–95). This can be seen as, in effect, one of the earliest climate-related modeling efforts.

In the subsequent Islamic tradition of Ibn Khaldûn, we find a more reflexive and social situating of knowledge. He invokes authority for his statements from "continued tradition," "personal observation," and the observations and practices of "many persons we used to know" (Ibn Khaldûn 1967: 54, 57, 68). Ibn Khaldûn (1967: 64) also gives us an explicit statement of the logic of the comparative method: "If one pays attention to this sort of thing in the various zones and countries, the influence of the varying quality of the climate upon the character of the inhabitants will become apparent."

Montesquieu, on the other hand, tests theories of climatic influence by changing the value of key social variables, for example, when members of an ethnic group move from one climate to another. He recognizes contradictions in some simplistic climatic explanations. Thus, Montesquieu (1752: 189, 191) admits that Indians seem to have a "very odd combination of fortitude and weakness." The key intellectual construct in his climate theory is based on his freezing and thawing of a sheep's tongue, and his observation of the attendant contracting and lengthening of its "papillae" (which is derivative of Hippocrates's own experiments freezing and thawing water two millennia earlier). Montesquieu argues by analogy that the human senses are dulled in cold climes and exaggerated in hot ones, which, he argues, helps to explain the difference in character between the two.

Zimmermann (1987: 96) maintains that the ancient Indian system followed an entirely different course from that of ancient Greece: "One difference of capital importance . . . affects the very nature of the *reasoning* in ancient India, the style of the Sanskrit texts and the rules of literary composition. Once a field, or subject of study has been determined, we, as the heirs of Greco-Latin logic, proceed by way of description and argument. . . . The situation in the Ayurvedic treatises, however, is quite different. In ancient

India, the primary material of knowledge was constituted by the *recitation of series of words* of a more or less stereotyped nature." Whereas the Greeks developed a system of natural history, in short, what the Indians developed was essentially a lexicon. As Zimmermann (1987: 97) writes, "While the Greek and Latin naturalists observed and registered curious details and invented the model of knowledge called *natural history*, the Indian scholars, following a quite different path of reasoning, forged and created a thesaurus of names and their semantic equivalents." Whereas the Greco-Latin system was evidence-based, the Indian one was based on a formal system of language. Zimmermann (1987: 9) writes, "Far from being an invention of man as a collection of empirical recipes would be, Ayurveda is beyond our powers of knowledge and submits us to laws which can be taught but cannot be discovered." Zimmermann (1987: 158) quotes one ancient Sanskrit text, the Susruta Samhita, as follows: "No need to examine them [the Ayurvedic treatises], no need to reflect on them, they will make themselves known, the remedies that the clear-sighted must prescribe in accordance with tradition."

One characteristic shared by both the Greek and Indian traditions is the synthesis of different areas of experience and knowledge. In the Indian case, climate is incorporated into a worldview that encompasses all life. As Zimmermann (1988: 205) writes, his analysis of the Indian tradition "blows up the so-called 'medical system,' to reach upstream the more basic categories of collective thought, and downstream the more basic facts of ecology." In the case of the Greek system, works like Hippocrates's "Airs, Waters, Places" reveal, as Glacken notes (1967: 84), "how closely interrelated are the early histories of medicine, geography, and anthropology." In both cases there are specific disciplines of knowledge that are accorded disproportionate importance. In ancient India it was diet, pharmacology, and what Zimmermann (1988: x) calls "the superior register" of physiology. As he writes (1987: 8), "What the Rishis, the seers of Vedic times, quite literally saw was that the universe is a kitchen, a kind of chemistry of *rasa*, diluted or sublimated to feed now one and then another of nature's kingdoms: the stars, the waters, the earth, the plants, the fauna."[2]

LIMITS TO KNOWLEDGE The nature of knowledge about climate—how it is acquired and by whom and by what methods—is a central focus of the contemporary debate about climate change (see O'Reilly, this volume). Indeed, even the possibility of such knowledge is debated, given what Hulme (2011: 266) calls "the voids in the human imagination" engendered by the fear and pessimism surrounding the topic. This crisis of confidence represents a break with the past. From the time of the Greeks down to, and reflected in, modern-day ecological theory, the idea of a designed earth is one of the central ideas in thinking about earth and man. Since, as Glacken (1967: 423) writes, "the wisdom of the Creator is self-evident," everything is potentially knowable, whether through religious doctrine or, increasingly, through history, scientific inquiry. Therefore, such a thing as knowledge about the earth and aspects of the earth such as its climates was thought to be possible—attainable through personal experience, effort, and talent.

Distinctions have commonly been made between the knowledge of those who study a matter and that of those who do not. As Ibn Khaldûn (1967: 65) says of the cultural-ecological differences of desert versus hill peoples, "This is known to those who have investigated." From ancient times, recognition of differences in knowledge is reflected in the identification of privileged sources, of expert witnesses. Ibn Khaldûn elsewhere avers that the health of chicks from hens fed on grain cooked in camel dung "has been mentioned by agricultural scholars and observed by men of experience" (1967: 69). Theophrastus (1926: 2:391) similarly writes, "The signs of rain, wind storm and fair weather we have described so far as was attainable, partly from our own observation, partly from the information of persons of credit." He goes on to list a half dozen named individuals known to him as "good astronomers" (1926: 2:393). Theophrastus also offers a remarkable early statement regarding the value of knowledge and its local situatedness: "Wherefore good heed must be taken to the local conditions of the region in which one is placed. It is indeed always possible to find such an observer, and the signs learnt from such persons are the most trustworthy" (1926: 2:393).[3]

The idea that there are multiple, alternate ways of knowing is a modern phenomenon. The Hippocratic corpus compares climates and races, but not *views* of climates and races. Whereas work by the likes of Theophrastus presents what is in effect a corpus of local knowledge of climate, it is not represented as such. It is not self-conscious ethno-science as we find in the few contemporary examples of ethno-meteorology (e.g., Puri 2007). For example, Sillitoe (1994: 246) introduces his study of the Wola of Papua New Guinea as follows: "It is an attempt at an ethno-meteorology, a scarcely researched but creditable field of enquiry. It compares and contrasts observations of the climate and records of the weather with local people's comments and thoughts on these phenomena." And Orlove et al. (2002) write of their study in the Andes, "To our knowledge, this is the first time a scientific explanation has been offered for the workings of a folk meteorological practice." Zimmermann's (1988: 198) comparison of the findings of modern natural science with the ancient Vedic treatises clearly falls within the same tradition of study as Orlove et al.: "My inquiry borrows illustrations from contemporary tropical geography, botany and zoology, but its point of departure is the study of texts. I am concerned with a corpus of traditional notions into which it is nevertheless possible to subsume observations from the natural sciences to discover what is at stake in Hindu scholasticism."

There is no such effort to reconcile folk and scientific views of climate in the case of the recent explosion in public rhetoric in the United States attacking the science of global climate change. The increasing repudiation of research on climate change (Smith and Leiserowitz 2012), which essentially amounts to an industrialized society's variant of folk theorizing about climate, is not being treated "ethnographically." Early views of open public debates about science as a modern product of the Enlightenment (Beck 1992) have given way to skepticism about the disinterestedness of such debate (Demeritt 2001), and some scholars have come to rethink long-standing academic practices of public critique of science.[4] As Bruno Latour (2004: 227) writes, "The danger would no longer be coming from an excessive confidence in ideological arguments posturing as matters of fact—as we have learned to combat so efficiently in the past—but from an excessive *distrust* of good matters of fact disguised as bad ideological biases! While we spent years trying to detect the real prejudices hidden behind the appearance of objective statements, do we now have to reveal the real objective and incontrovertible facts hidden behind the *illusion* of prejudices?"

Nature and Culture

An enduring theme in the history of thinking about climate and society is the conception of nature and culture and the relationship between them. Hippocrates sought to find the answer to human difference in differences in nature. Dramatic differences in nature were thought to produce dramatic differences in human beings and their societies. It is thus culture that is problematized and explained in this formulation, not nature; nature is the independent variable, and culture is the dependent one. Explanation of difference in the world was therefore sought by examining the impact of nature on culture, not the reverse. Glacken (1967: 121) notes the failure of the classical scholars to examine the influence of culture on nature and suggests that this has had long-lasting consequences: "If Plato in the *Laws* had noted that men change their environments through long settlement and that soil erosion and deforestation are parts of cultural history, he could have introduced at an early time these vital ideas into cultural history and changed the course of speculation regarding both man and environment." What efforts there were to study the influence of culture on nature were distinct from those focusing on the reverse—the circle was not completed. There were no reciprocal studies (Glacken 1967: viii).

The classical emphasis on nature's determination of culture has influenced millennia of thought. Montesquieu (1752: 194) clearly follows Hippocrates in locating causality in nature: "It is the variety of wants in different climates that first occasioned a difference in the manner of living, and this gave rise to a variety of laws." Several centuries prior, Ibn Khaldûn (1958: 1:175) was likewise interested in how variation in nature explains variation in culture, as he recommends studying "the influence of the varying quality of the climate upon the character of the inhabitants." As a result, a remarkably constant philosophy of nature's determinism has dominated Western thinking for more than two millennia (Glacken 1967: 81): "Virtually all the familiar assertions of modern times, even if we charitably stop with Montesquieu, are found in cruder form in antiquity: warm climates produce passionate natures; cold, bodily strength and endurance; temperate climates, intellectual superiority; and among the non-physiological theories, a fertile soil produces soft people, a barren one makes one brave."

Although enlightenment thinkers like Voltaire critiqued simplistic notions of environmental determinism, these ideas were still robust in the late nineteenth and early twentieth centuries, as reflected in the works of Friedrich Ratzel and his followers, like Ellen Churchill Semple and Ellsworth Huntington. The excesses of this position led to a reaction, such that for a long time in the twentieth century historians and social scientists simply ignored environment altogether. Kroeber (1947: 3) acknowledges but also laments this abandoning of the environment: "For a generation American anthropologists have given less and less attention to environmental factors. In part this represents a healthy reaction against the older naive view that culture could be 'explained' or derived from the environment."

Celebration versus diminution of human agency with respect to the environment seems to be a modern idea (Dove et al. 2008). The idea of humans as geographic or geological agents (Chakrabarty 2009) is a modern development. Glacken (1967: 710) suggests that it first appeared in eighteenth-century natural histories. The nineteenth-century work by Marsh (1865) was also a pivotal development of the idea of human impact on nature. This idea was further refined in the field of environmental history in the late twentieth century, which took a balanced, reciprocal view of nature–culture relations, as reflected in the work of Ladurie (1974), Worster (1979), Cronon (1983), and Grove (1995), among others.

One would have expected this trend to continue and even be strengthened in the twenty-first century, given the amount of human agency that would appear to be reflected in global climate change and the development of the aforementioned concept of the Anthropocene. The picture is more complicated, however. Whereas scholars like Chakrabarty (2009) see modern climate change as elevating humans to the status of geological actors, others, like Hulme (2011), see climate change science as reintroducing environmental overdetermination of human fates. Hulme (2011: 250) sees an asymmetry in representations of future climate change and social change (cf. Demeritt 2001: 313, 316, 318–19). Because of a bias in favor of quantitative, predictive models, the natural and biological sciences have acquired hegemony over visions of the future, which inevitably leads to the downplaying of human agency. Hulme writes, "These models and calculations allow for little human agency, little recognition of evolving, adapting, and innovating societies, and little endeavor to consider the changing values, cultures, and practices of humanity" (2011: 255–56). A diminished view of human agency has led to an overdetermining of the future in terms of the environment. Hulme concludes, "And so the future is reduced to climate. By stripping the future of much of its social, cultural, or political dynamism, climate reductionism renders the future free of visions, ideologies, and values. The future becomes overdetermined" (2011: 264–65). He calls this "climate reductionism" or "neoenvironmental determinism" (Hulme 2011: 247).

The Other

Overlapping these ideas of nature are ideas about culture, about people (Williams 1980: 67). Hippocrates (1923: 111) writes, "The races that differ but little from one another I will omit, and describe the condition only of those which differ greatly, whether it be through nature or through custom." One of the two or three most important points that Glacken (1967) makes in *Traces on the Rhodian Shore* is to highlight the foregoing passage from Hippocrates and its vast historical consequences. He writes, "Hippocrates says that he is not concerned with peoples that are similar but with those that differ, either through nature or custom from one another. This matter-of-fact statement may reveal the reason for the interest in differing physical environments in antiquity. . . . If Hippocrates had shown an interest in accounting for similarities rather than differences, the history of environmental theories would have been entirely different" (Glacken 1967: 85). The classical explanation of different peoples in terms of different climates was a search for an answer to the age-old question: Why is the other different? Hippocrates could have seen like people in unlike environments as posing a question—namely, Why are people *the same?*—but he did not. The latter question would have been an equally logical one to ask; that it was not asked had to do with a fundamental bias in the way questions about nature and culture were framed.

Hippocrates' approach to this matter is didactic; his intent is "to show" human diversity: "Now I intend to compare Asia and Europe, and to show how they differ in every respect, and how the nations of the one differ entirely in physique from those of the other" (1923: 105). His intent is not to present an agenda for action. Given the perceived influence of nature upon culture, as opposed to the reverse, there is no real possibility of action, of change. This argument offers a ready explanation of human difference, in particular problematic difference. As Glacken (1967: 258) writes, this explanation is "serviceable in accounting for cultural, and especially for racial, differences"; and thereby it helps to justify privilege. It is no coincidence that from the time of Montesquieu environmental determinism and colonial and postcolonial political-economic projects have had a shared genealogy. The continued power of such explanations can be seen in the current popularity of works of global geographic determinism (e.g., Diamond 1997, 2005). Less obviously, the continuing influence of this tradition of explanation can perhaps also be seen in the self-privileging stances being taken by the industrialized nations toward the late-industrializing nations, which are less responsible for but more vulnerable to climate change (Dove 1994; Lahsen 2004), as well as in the equally self-privileging distinctions being made between more developed countries with high "adaptive capacity" versus less-developed countries with perceived low capacity (Moore 2011).

Although Hippocrates and those who followed him drew their lessons about difference from peopled landscapes, the anthropogenic character of those landscapes played no part in these lessons. The dimensions of the environment selected for emphasis in these historic texts were climatic hot/cold, wet/dry—dimensions that were then beyond human influence. Thus nonhuman phenomena were invoked to explain human difference. Today we can imagine it otherwise. We can imagine anthropogenic environmental characteristics being invoked to explain differences among humans. Desertification, deforestation, and indeed climate change are all invoked in contemporary environmental discourses to explain why diverse societies are impoverished, violent, and so on.

Today we also draw lessons from unpeopled landscapes. The modern imagination, at least in the industrialized West, has been captured by the idea of empty nature. This is reflected not only in the enormous modern interest in wilderness but even in climate change discourse, which, notwithstanding its anthropogenic drivers, accords special status to the Arctic, alpine glaciers, and tiny islands (see Orlove et al. in this volume). All of this represents a departure from the historic texts discussed here. As just noted, the exotic *other* landscape of the historic imagination was nearly always a peopled one. An empty nature seemingly served no pedagogic purpose in these texts, unlike the modern case. Unpeopled landscapes had no pedagogic value for explaining difference among humans.

In these early eras there was more unpeopled landscape on the earth than peopled. In a much less peopled world, the empty waste bore no lesson. Empty nature was not needed as a foil until the modern era, when it is nearly gone, and it takes an act of imagination to see it, to construct it. There has been, therefore, a historic reversal in this geography of pedagogy. Some environmental scholars do not see this as a salutary development, based on a critique of the role of the so-called wild in modern environmental thinking. As Cronon (1996: 81) writes, "To the extent that we live in an urban-industrial civilization but at the same time pretend to ourselves that our *real* home is in the wilderness, to just that extent we give ourselves permission to evade responsibility for the lives we actually lead." In short, Cronon sees the modern pedagogical value of the wild, the waste, as negative.

Climate determinists, from the earliest scholarly traditions discussed here to the modern era, have invariably located the climatic and societal ideal in their own latitude and the climatic and societal opposite somewhere else. To serve the purpose of the comparison, the *other* cannot be too distant, either geographically or conceptually. Hippocrates's question as to why the *other* is different can be asked and answered only if there is an assumption of an underlying common humanity. The *other* is different but still human. Indeed, it might be more correct to call the *other* at the other latitude a double, someone a little too close for comfort, which leads to defamiliarization. With greater difference, however, the basis for comparison breaks down. There was neither a felt need nor an actual effort to explain in terms of climate the difference of humans from the fabled races of antiquity, like the Cyclops, for example. It was only the existence of *others* like us, yet unlike us in *some* respects, that posed the question, Whence comes difference? The existence of *others* either completely like or completely unlike humans posed no question at all.

The *other* or double also plays an important role in contemporary climate change discourse. For example, Farbotko and Lazrus (2012) have studied the characterization of the people of Tuvalu as "climate refugees," which they liken to glaciers and polar bears as an exemplary category in climate change discourse. They argue that refugees, glaciers, and polar bears are employed either implicitly or explicitly in metaphors like "the laboratory," "the litmus test," and "the canary in the coal mine," to warn other parts of the world about the future threat that climate change poses to them. Refugees, glaciers, and polar bears are distant from the global center but serve it through a process that Farbotko and Lazrus liken to ventriloquism, in that a threat to the other is reinterpreted as a threat to the self.

As in the Vedic, Greek, and Islamic traditions, therefore, contemporary climate change discourse as critiqued by Farbotko and Lazrus removes the environmental problematic from us and sets it at a distance (Smith and Leiserowitz 2012: 1023). However, running through modern environmental theory, as in Cronon's work, is a warning against doing just this. Hewitt (1983) argues that such removal is often a partisan act, leads to an incomplete view of the problem, and usually does not lead to effective solutions. For analogous reasons, some scholars think that apocalyptic rhetoric about climate change is ultimately unhelpful. As Keller (1999) writes, "Apocalyptic habits . . . are not conducive to long-term and sustainable life on this planet."

THE VALUE OF THE COMPARATIVE STUDY OF HISTORY

The *other*, and indeed much of what has been discussed in this chapter, is based on a process of comparison. This is also what we are doing when we examine historic studies of climate and society, when we respond to the enduring interest in the association of hot/cold and identity to think across eras. This discussion of historic studies of climate and society reveals them to be both accessible and inaccessible: they are familiar enough to be accessible but foreign enough to make for interesting comparison. I have suggested that these historic studies can be termed *doubles*, like Hippocrates's (1923: 109) "wild and uneven" land, "where the seasons experience the most violent and the most frequent changes." Comparative study, thinking across cultures, has been an integral dimension of anthropology since the birth of the discipline, and it is an important part of what anthropology has to contribute to modern studies of climate change.

What does this sort of contrast and comparison of like and unlike accomplish? These historic texts are valuable not only because they illuminate present thinking about climate with their similarities but also because they decenter our present thinking by virtue of their dissimilarities—which are potent only because they are coupled with similarities. Historic studies thereby defamiliarize the present and can illuminate it in a unique way. After millennia of asking, Why are we different?, in short, we can ask, Why and how are we the same as well as different? This has been perhaps the central question in anthropology since its inception.

When we trace continuities in this manner, when we make linkages, when we combine, we are creating something new. Such comparisons are not givens and therefore they are not a priori constructions. To compare is to make something that was not there before. In so doing, we put modern thought in a new light. The very idea of a historic, ancient comparator gives us a fresh perspective on contemporary thinking about climate and society. The historicizing of contemporary thought offers an alternative (possibly a productive one) to saying that modern climate change represents the end of history. At least, this deep historical perspective makes any such claim much more challenging. This critique is very much part of the anthropological project, which has always attended not just to history but to the very idea of history.

NOTES

1. Related developments in anthropology involving the emergence of "multi-species ethnography" (Helmreich and Kirksey 2010) reflect a recent effort to consciously problematize the human / nonhuman conceptual divide (Ingold 2000).
2. There is a privileged register in the contemporary discourse of climate change: predictive models from the natural sciences. As Hulme (2011: 264) writes, "Although it is clear to many scientists that 'the impact of any climatic event depends on the local ecological setting and the organisational complexity, scale, ideology, technology and social values of the local population,' current intellectual endeavors in this area unduly privilege climate as the chief determinant of humanity's social futures."

3. Just as there *are* some persons who are more trustworthy than others, so are there those who are less trustworthy; that is, the absence of knowledge, and incorrect versus correct views, also are recognized. Thus Ibn Khaldûn (1958: 104, 107–8) offers to "explain and prove" the views of "the philosophers" that the "excessive heat and slightness of the sun's deviation from the zenith" make human society untenable south of the equator, partly agreeing and partly disagreeing with them. He flatly takes issue with "the assumptions of physicians" regarding the effects of hunger on the human constitution (1958: 181–82).
4. Critique of science continues, even in the field of climate change, as exemplified by Hulme (2011).

REFERENCES CITED

Aristotle. 1952. *Meteorologica.* Translated by H. D. P. Lee. Cambridge: Harvard University Press.

Beck, U. 1992. *Risk Society: Towards a New Modernity.* Translated by M. Ritter. London: Sage Publications.

Chakrabarty, D. 2009. "The Climate of History: Four Theses." *Critical Inquiry* 35: 197–222.

Cronon, W. 1983. *Changes in the Land: Indians, Colonists, and the Ecology of New England.* New York: Hill and Wang.

———. 1996. "The Trouble with Wilderness, or Getting Back to the Wrong Nature." In *Uncommon Ground: Rethinking the Human Place in Nature,* ed. William Cronon, 69–90. New York: W. W. Norton.

Crutzen, P. 2002. "Geology of Mankind." *Nature* 415: 23.

Demeritt, D. 2001. "The Construction of Global Warming and the Politics of Science." *Annals of the Association of American Geographers* 91, no. 2: 307–37.

Derrida, Jacques. 1994. *Spectres of Marx.* New York: Routledge.

Diamond, J. M. 1997. *Guns, Germs, and Steel: The Fates of Human Societies.* New York: W. W. Norton.

———. 2005. *Collapse: How Societies Choose to Fail or Succeed.* New York: Viking.

Dove, M. R. 1994. "North–South Relations, Global Warming, and the Global System." Special issue on "Human Impacts on the Pre-Industrial Environment." *Chemosphere* 29, no. 5: 1063–77.

Dove, M. R., A. Mathews, K. Maxwell, J. Padwe, and A. Rademacher. 2008. "The Concept of Human Agency in Contemporary Conservation and Development." In *Against the Grain: The Vayda Tradition in Human Ecology and Ecological Anthropology,* ed. B. Walters, B. J. McCay, P. West, and S. Lees, 225–53. Lanham, Md.: Lexington Books.

Farbotko, C., and H. Lazrus. 2012. "The First Climate Refugees? Contesting Global Narratives of Climate Change in Tuvalu." *Global Environmental Change* 22: 382–90.

Glacken, C. J. 1967. *Traces on the Rhodian Shore: Nature and Culture in Western Thought from Ancient Times to the End of the Eighteenth Century.* Berkeley: University of California Press.

Grove, R. H. 1995. *Green Imperialism: Colonial Expansion, Tropical Island Edens, and the Origins of Environmentalism, 1600–1860.* Cambridge: Cambridge University Press.

Helmreich, S., and E. Kirksey. 2010. "The Emergence of Multi-Species Ethnography." *Cultural Anthropology* 25, no. 4: 545–76.

Hesiod. 1914. *Works and Days.* Translated by G. Evelyn-White. Gloucester, UK: Dodo Press.

Hewitt, K. 1983. "The Idea of Calamity in a Technocratic Age." In *Interpretations of Calamity, From the Viewpoint of Human Ecology*, ed. K. Hewitt, 3–32. Boston: Allen and Unwin.

Hippocrates. 1923. "Airs, Waters, Places." In *Hippocrates.* Volume I. Translated by W. H. S. Jones, 65–137. Cambridge: Harvard University Press.

Hulme, M. 2011. "Reducing the Future to Climate: A Story of Climate Determinism and Reductionism." *Osiris* 26: 245–66.

Ibn Khaldûn. 1958. *The Muqaddimah: An Introduction to History*, 3 vols. Translated by Franz Rosenthal. Princeton: Princeton University Press.

Ingold, T. 1993. "Globes and Spheres: The Topology of Environmentalism." In *Environmentalism: The View from Anthropology*, ed. Kay Milton, 31–42. London: Routledge.

———. 2000. *The Perception of the Environment: Essays on Livelihood, Dwelling and Skill.* London: Routledge.

Keller, C. 1999. "The Heat Is On: Apocalyptic Rhetoric and Climate Change." *Ecotheology* 7: 40–59.

Kroeber, A. L. 1947. "The History of Concepts." In *Cultural and Natural Areas of Native North America*, 3–12. Berkeley: University of California Press.

Ladurie, E. L. R. 1974. *The Peasants of Languedoc.* Translated by John Day. Urbana: University of Illinois Press.

Lahsen, M. 2004. "Transnational Locals: Brazilian Experience of the Climate Regime." In *Earthly Politics: Local and Global in Environmental Governance*, ed. Sheila Jasanoff and Marybeth L. Martello, 151–72. Cambridge: MIT Press.

Latour, B. 2004. "Why Has Critique Run Out of Steam? From Matters of Fact to Matters of Concern." *Critical Inquiry* 30, no. 2: 225–48.

Launay, R. 2010. *Foundations of Anthropological Theory: From Classical Antiquity to Early Modern Europe.* Malden, Mass.: Wiley-Blackwell.

Lawrence, B. B. 2005. "New Introduction." *Ibn Khaldûn. The Muqaddimah: An Introduction to History.* Abridged, edited, and translated by Franz Rosenthal. Princeton: Princeton University Press.

Levin, L. M. 1936. *The Political Doctrine of Montesquieu's Esprit des Lois; Its Classical Background.* New York: Institute of French Studies, Columbia University.

McIntosh, R. 2000. "Social Memory in Mande." In *The Way the Wind Blows: Climate, History and Human Action*, ed. Roderick J. McIntosh et al., 141–80. New York: Columbia University Press.

Marsh, G. P. 1965. *Man and Nature: Or, Physical Geography as Modified by Human Action.* Cambridge: Harvard University Press.

Montesquieu, Charles de Secondat, baron de. 1752. "Book XIV of Laws in Relation to the Nature of the Climate." In *The Spirit of the Laws (Esprit des loix)*, translated by Thomas Nugent. London: J. Nourse and P. Vaillant.

Moore, F. C. 2011. "Costing Adaptation: Revealing Tensions in the Normative Basis of Adaptation Policy in Adaptation Cost Estimates." *Science, Technology, and Human Values.* Published online March 23.

Neuman, F. 1949. *Introduction. The Spirit of the Laws, by Baron de Montesquieu.* Translated by T. Nugent. New York: Harper.

Orlove, B. S., John C. H. Chiang, and Mark A. Cane. 2002. "Ethnoclimatology in the Andes." *American Scientist* 90: 428–35.

Pliny. 1942. *Natural History.* 10 vols. Translated by H. Rackham. Cambridge: Harvard University Press.

Puri, R. K. 2007. "Responses to Medium-Term Stability in Climate: El Niño, Droughts and Coping Mechanisms of Foragers and Farmers in Borneo." In *Modern Crises and Traditional Strategies: Local Ecological Knowledge in Island Southeast Asia,* ed. R. F. Ellen, 46–83. New York: Berghahn Books.

Rosenthal, F. 1958. "Translator's Introduction." In *The Muqaddimah: An Introduction to History.* 3 vols. by Ibn Khaldûn. Princeton: Princeton University Press.

Sider, D., and C. Brunschön, eds. 2007. *Theophrastus of Eresus: On Weather Signs.* Leiden: Brill.

Sillitoe, P. 1994. "Whether Rain or Shine: Weather Regimes from a New Guinea Perspective." *Oceania* 64: 246–70.

Smith, N., and A. Leiserowitz. 2012. "The Rise of Global Warming Skepticism: Exploring Affective Image Associations in the United States Over Time." *Risk Analysis* 32, no. 6: 1021–32.

Theophrastus. 1926. *Enquiry Into Plants and Minor Works on Odours and Weather Signs.* 2 vols. Translated by Arthur Hort. Cambridge: Harvard University Press.

Toynbee, A. J. 1935. *A Study of History.* 3 vols. London: Oxford University Press.

Virgil. 2004. *Georgics.* Translated by Peter Fallon. Oxford: Oxford University Press.

Weiss, H. 2000. "Beyond the Younger Dryas: Collapse as Adaptation to Abrupt Climate Change in Ancient West Asia and the Eastern Mediterranean." In *Environmental Disaster and the Archaeology of Human Response,* ed. G. Bawden and R. Reycraft, 75–95. University of New Mexico, Anthropological Papers 7.

Williams, R. 1980. *Problems in Materialism and Culture: Selected Essays.* London: NLB.

Wilson, E. O. 1996. *In Search of Nature.* Washington, D.C.: Island Press.

Wood, D. 2006. "On Being Haunted by the Future." *Research in Phenomenology* 36, no. 1: 274–98.

Worster, D. 1979. *Dust Bowl: The Southern Plains in the 1930s.* New York: Oxford University Press.

Zalasiewicz, J., M. Williams, W. Steffen, and P. Crutzen. 2010. "The New World of the Anthropocene 1." *Environmental Science and Technology* 44, no. 7: 2228–31.

Zimmermann, F. 1987. *The Jungle and the Aroma of Meats: An Ecological Theme in Hindu Medicine.* Berkeley: University of California Press.

———. 1988. "The Jungle and the Aroma of Meats: An Ecological Theme in Hindu Medicine." *Social Science and Medicine* 27: 197–206.

Chapter 2 **How Long-Standing Debates Have Shaped Recent Climate Change Discourses**

Ben Orlove, Heather Lazrus, Grete K. Hovelsrud, and Alessandra Giannini

In the past decade or two climate change impacts have been presented as a pressing concern in some places, while in others they seem less urgent. The melting of the world's ice caps and Arctic sea ice and the associated rise in sea level that may inundate low-lying islands are recognized as harmful consequences of a changing global climate, while other trends—for example, deterioration of mountain ecosystems, the shifts in precipitation and vegetation that cause deserts to spread—are not seen as being serious or as closely linked to climate change. We examine the discussions of place-based climate impacts and emphasize two characteristics of this conversation. First, we see it as an extension of earlier conversations—some of them from previous centuries—about the nature of places and the nature of the earth. Second, we see it as involving a number of speakers in the centers and at the margins of world power. We pay particular attention to the interactions between the European and American colonizers and the colonized peoples, and, granted our long time horizon, between the antecedents and successors of these. We engage with the global conversation about climate change impacts through an examination of four regions, namely, the Arctic, islands, deserts, and mountains. We focus on these four regions and their histories to illustrate the long historical depth of discourses that shape current climate change discussions and to show that these discourses are not the product of the representatives of powerful nations alone but also represent the engagement of weak, often disenfranchised speakers from the regions distant from centers of power.

We note that our claims are contested. Our first claim is of the long history of the discussions that shape the particular treatment of different places in climate change discourse. Others who have discussed climate change frameworks have stressed their short historical depth. They often associate these frameworks with recent forms of scientific practice, such as complex numerical models which require global networks of instrumentation and advanced computational capacity to run simulations of the earth's climate. Demeritt (2001) stresses the global scale of climate change and the importance of general circulation models within the geosciences for framing the problem. Swyngedouw (2010) presents this use of models and technocratic political discourses, also a development of recent decades, as the twin supports of present-day climate change frameworks. Our second claim, the view that subordinate peoples have taken part in the construction of climate change discourse, is also contested. Others see them as separated from it or as recipients of it rather than participants in it. Jasanoff argues that there is a cosmopolitan discourse of climate change emanating from powerful nations and international organizations. She states, "Climate . . . is everywhere and nowhere, and hence not easily accessible to imaginations rooted in specific places" (Jasanoff 2010: 237). She continues, "Climate change . . . can be linked to a place, but that place is the whole Earth" (2010: 241). Gupta (1998) and others stress the power of new, globally unifying discourses, particularly environmentalism and sustainability, which inspire action around the world but result in erasing earlier, often more local, frameworks. These views resonate strongly in the present moment in anthropology, in which an attention to universalisms (human rights, science, development, and so forth) has replaced an earlier interest in globalization (Tsing 2005).

In questioning the recency and univocality of climate change discourse we acknowledge that the human influence on the climate is recent and of a planetary scale (Crutzen 2006; Latour 2004). Greenhouse gas emissions that diffuse through the atmosphere from burning fossil fuels are long-lived, hence well mixed. Greenhouse gases alter the balance between the solar energy reaching the earth and the energy radiating back out into space, so that efforts to address climate change—adaptation as well as mitigation—must include the coordinated efforts of many, if not all, nations. We also recognize that many consequences of climate change, such as the pressures on food production, are worldwide in scale.[1] Nonetheless, we argue that specific places rather than the planet as a whole are emphasized in public and scientific understandings and debates about anthropogenic climate change.

We trace the historical depth and dialogic nature of climate change discourses by looking closely at the Arctic, islands, mountains, and deserts, where recent climate impacts have been evident. In three cases, climate change makes areas, or portions of them, vulnerable to disappearance (low-lying islands, glaciers) or to loss of fundamental characteristics (the Arctic). In the final case, deserts, the risk is expansion, specifically the disappearance of areas with extensive vegetation and their replacement by more barren zones.

We emphasize four phases of the historical development of current frameworks. First, each of the encounters between Europe and the four regions begins before the major European expansion and even before the Age of Discovery, conventionally associated with the fifteenth century, which saw Columbus's landfall in America at its close. Second, the regions were the targets of attention during the period of colonial expansion and rule. This attention stemmed from material interests, whether directly for the extraction of economically important resources or indirectly for geopolitical and strategic significance, but the regions and their indigenous peoples also appealed deeply to the cultural imagination of colonial travelers and scientists. These material and cultural forces contribute to the positioning of the regions in relation to climate change. Third, in the postwar period, the regions (some as newly independent nations, some as regions of older nations) gained visibility as key exemplars of environmental discourses and projects. Two regions, deserts and mountains, emerged as leading examples of environmental fragility, understood as destructive land-use practices on a local scale; the other two regions, the Arctic and low-lying islands, emerged as examples of global environmental injustice, understood as pollution or resource exploitation on a global scale. In the final period, starting in the mid-1980s, the Arctic and low-lying islands, building on identities and organizations that crystallized in the first decades of the postwar period, became closely associated with climate change, while deserts and mountains, retaining a concern with local land use, did not. Instead, the latter remained associated with sustainable development, an environmental framework which, though linked to climate change, is different from it in significant ways.

These four phases can be found in each of the core regions, though their temporal boundaries vary somewhat; the timing of their start and close depend on dynamics emerging both in the centers of power and in the specific regions. To present the historical narrative in an effective way, we group together the first two phases and the second two phases.

THE HISTORICAL ORIGINS AND DEVELOPMENT OF CLIMATE CHANGE DISCOURSES

Period 1: Exploration and Colonization

As Wolf argues in his *Europe and the People Without History* (1982), the history of European colonization was preceded by centuries in which areas distant from the core regions of the emerging European states were known through a variety of sources, including myths, scattered travel accounts, and linkages through long-distance trade. These images were reshaped by a number of forces, among them the economic motives to extract resources and control strategic areas and cultural impulses to construct scientific and artistic representations. The dominant cultural frameworks at the beginning of this period were strongly influenced by Christian religious beliefs and the associated political ideologies of monarchies; these provided additional justification for the exploration and expansion into areas inhabited by non-Christian people. Major cultural and intellectual shifts within Europe from the seventeenth century onward influenced the perception of other regions as well (see Dove, this volume). To pick just a few of the most important tendencies, the Enlightenment from the eighteenth century on supported an insatiable curiosity of people to collect and categorize a wide variety of phenomena. The romantic movement of the nineteenth century led to a fascination with exotic cultures, which were treated as coherent wholes to be experienced and savored. The interactions of these two tendencies led both to a firm belief in the progress of individual nations and of humanity at large and to a move toward wider political participation within such progress. This variety of sources and motives characterizes all four regions.

In the Arctic, the remote regions of high latitudes, settled millennia ago by hunter-gatherers, were the subject of ancient myth but little visited by outsiders until the Viking expansion from southern Scandinavia in the ninth and tenth centuries during the relatively warm "medieval climatic optimum" (Mann 2002). As a result of this movement, European settlements were established in northern Scandinavia, Iceland, and Greenland. This expansion brought far northern regions and their products, such as fish, furs, and ivory, into closer contact with Europe. There was retreat and abandonment of the northernmost Viking settlements in Greenland after a shift to colder climates during the Little Ice Age, population declines after the Black Death, and competition from African ivory.

With the growth of trade in the early modern period, the European powers competed to discover sea routes from Europe to Asia across the top of North America through the Northwest Passage and over Europe and Siberia in the Northeast Passage.[2] The Dutch reached Svalbard, north of Norway, in 1596. Russians, traveling close to the northern shore of Asia, arrived in the Pacific in the seventeenth century. They continued on to Alaska, exploring it in the 1740s and establishing settlements soon thereafter in the Aleutian chain. This expansion involved interactions with indigenous populations, who provided food and served as guides.

The search for the Northwest Passage led to early exploration of the eastern coast of North America in the sixteenth and seventeenth centuries. The thick sea ice of the high Arctic was a major obstacle for explorers. Nonetheless, exploration of the high Arctic was active. Whaling became an important commercial activity in the seventeenth century and continued well into the nineteenth century. Sizable coal deposits led to permanent settlements in Svalbard. It was not until the first half of the nineteenth century, though, that the Canadian Arctic was well explored and the early twentieth century that the North Pole was reached. These expeditions were often conducted with the aid of indigenous peoples, whose familiarity with the Arctic environment was crucial. By the end of the nineteenth century all portions of the Arctic had been incorporated into the territories of the nations adjacent to them.

In addition to economic and geopolitical interests, scientific exploration developed in the late nineteenth century. A number of countries sent their own expeditions and collaborated on the organization of major scientific conferences. The first was the International Polar Year (1882–83), which was initiated by Austro-Hungarian explorers and also involved the United States, Canada, Russia, Germany, and other European countries. An intergovernmental organization of a sort, it concentrated predominantly on the Arctic, devoting less effort to the Antarctic, and it focused solely on the natural sciences. A number of research stations were set up and staffed for periods of months or years. The researchers recorded data on weather, ice, and ocean conditions and observed the aurora borealis and the earth's magnetic field. This effort advanced the Arctic as an object of study and also established a set of nations involved in the region. An International Polar Year has been held roughly every fifty years, the latest taking place in 2007–8, in which for the first time social scientists and Arctic indigenous peoples were fully included in research efforts.

If the Arctic was the object of international science and international conferences, it also held a broad cultural fascination. Its wildness was of great appeal to romantic painters and writers. Eskimo exhibits, as they were then called, were popular at the 1893 and 1904 World's Fairs. Native artifacts from the Arctic were prized in collections of ethnographic and natural history museums across North America and Europe. The film *Nanook of the North* (1922) attracted wide audiences. The iconic cartoons of the Arctic, often containing Inuit and igloos in a landscape of snow and ice, attest to the presence of the Arctic in popular culture (fig. 2.1). There was a parallel fascination

"I can get the latest market news from the New York stock exchange."

2.1: Arctic cartoon, with Eskimos and igloo by Joseph Farris. www.CartoonStock.com.

with the Sámi of Scandinavia, peoples whose traditional lifestyles centered on fishing, reindeer hunting and herding, and some agriculture.

As in the case of the Arctic, the discussion of low-lying islands dates back centuries before the major waves of European colonial expansion. Medieval Europeans knew of the Canary Islands off Africa, which had been long settled, and probably of the Azores and Madeira; their stories echoed a long-standing fascination with the remoteness of islands and bore traces of Greek myths of Atlantis and accounts of Arab travelers (Gillis 2004). Europeans visited these Atlantic islands in the fourteenth century and settled them in the early fifteenth century, displacing the native populations of the Canaries and establishing vineyards and plantations of sugar cane. This momentum continued with the discovery and conquest of native societies in the Caribbean from the end of the fifteenth century onward and with Spanish exploration of the Pacific in the sixteenth and seventeenth centuries. The Spanish controlled the Caribbean but were seriously challenged by the British and French beginning in the mid-seventeenth century; the region had highly productive slave-run plantations. The Indian Ocean, which had a long history of trade dominated by Arabs, also received European explorers, first Portuguese and then Dutch, French, and English, who conquered the few large islands in the region. The low-lying Maldives, long settled by South Indian peoples, were an Islamic sultanate that became a British protectorate in the late nineteenth century. Major European exploration of the Pacific began in the eighteenth century. The French and British concentrated on the larger, more mountainous islands, which offered more resources and greater possibilities of establishing colonies. They valued these islands for their geopolitical importance in supporting commercial and military control and for economic production of plantation crops; they were fascinated with the chiefdoms and kingdoms they encountered and with the traditions of interisland travel and exchange as well (Grove 1995).

The explorers were long aware of low-lying islands but did not direct their efforts toward them. The word *atoll*, which comes from the Maldivian language, first appeared in print in 1625 but was not common until the mid-nineteenth century. Visits to atolls began in the first half of the nineteenth century, when imperial competition was strong, as French expeditions arrived in the late 1820s. Charles Darwin stopped at a number of atolls in the Pacific and eastern Indian Oceans during the voyages of the *Beagle* in 1831–36, and the United States, developing a presence in the Pacific, conducted the Exploring Expedition of 1838–42. These travelers gathered information on local populations and collected specimens for natural history and ethnographic museums. Atolls attracted some attention from scientists (Darwin correctly explained their formation on the basis of observations he made during his voyages), but, unlike the Arctic, they were not constituted as a major subject of scientific research. The first notable international organization founded for the study of coral reefs, which are associated with atolls, is the International Society for Reef Studies, not founded till 1980.

Much like the Arctic, tropical islands have long been prominent in world culture, sometimes seen as benign paradises with balmy climates and pleasant, peaceful inhabitants and at other times as dangerous places filled with diseases and savage, even cannibalistic, natives (Grove 1995). They are now widely familiar as tourist destinations and as the subject of countless films. Like the Arctic, these small islands are often represented in cartoons showing a few castaways sitting under a palm tree on a tiny bit of sand in an immense ocean (fig. 2.2).

2.2: Desert island cartoon, with castaways and palm tree by Mike Gruhn. Gruhn at WebDonuts.com.

Generally lacking major geopolitical import and resources that could be exploited commercially, low-lying islands initially attracted little attention from imperial powers. Some of the last places to be claimed by these powers were atolls, such as Tuvalu. Though the Spanish islands in the Caribbean drew attacks from other colonial powers, large portions of Micronesia remained under Spanish rule until the late nineteenth century, when Germans, Japanese, and, later, the United States gained control of them. The United States established the Guano Islands Act in 1854, laying claim to uninhabited islands that contained guano, a widely used fertilizer, and took possession of some atolls in the Pacific and a few small islands in the Caribbean (Cushman 2013). Minor colonies and possessions through World War II and into the 1950s, the islands in the Pacific gained independence in the sixties and seventies, forming a number of separate countries. They established the South Pacific Forum in 1971 to promote cooperation. Atolls elsewhere had a similar history; for example, the Maldives, a British protectorate since the 1880s, became independent in the 1960s.

Mountain regions offer a different history. Bronze Age materials found in the Alps show that Europeans have traveled across high passes for millennia. Herders, hunters, miners, and foresters in medieval times would visit high mountain regions as well, though they and others saw them as harsh, dangerous areas (Nicolson 1959). Travelers who crossed mountain passes avoided glaciers, and local people dreaded their downslope advances, which would cover pastures, forests, and farms, particularly during the Little Ice Age from the mid-sixteenth to the mid-nineteenth centuries.

Mountain climbing developed slowly in Europe. The often-cited beginning was the ascent of Mont Ventoux in France by the Italian poet Petrarch in 1336, an early expression of Renaissance interest in empirical observation and curiosity about the natural world. Other ascents took place sporadically in following centuries. Inspired by the romantic fascination with wild nature and intense personal experience, a wave of ascents of glacierized peaks began in the eighteenth century in the Alps and in the first decades of the nineteenth century in Norway; these relied heavily on local guides and porters. A group of British mountaineers founded the Alpine Club in 1857 and climbed actively, ascending the last major unclimbed peak in the Alps, the Matterhorn, in 1865. They and other mountaineers moved on to the Pyrenees and the Caucasus and to high mountains around the world, a trend that culminated in the ascent of Everest by Edmund Hillary and a Sherpa guide, Tenzing Norgay, in 1953. The Alps have remained at the center of sport climbing internationally.

The travels of European climbers, especially in Asia, led to encounters with non-Western traditions that emphasized the spiritual quality of mountains as realms of purity and enlightenment. Exploration has been paralleled by a broad cultural interest in mountains, with extensive tourism and recreation in mountain regions. Like the other cases, mountains have long been important in world culture; they are seen as sublime, inspiring places, sites of pilgrimage and renewal and the preferred destination of poets such as William Wordsworth. Like the Arctic and small islands, mountains are featured prominently in cartoons, often showing a lone climber who reaches a cave inhabited by a wise old hermit—a partially ironic reference to the notion of mountains as spiritual (fig. 2.3).

2.3: Mountain cartoon, with guru and cave. © 2008 Mark Parisi. www.offthemark.com.

Overlapping the general cultural appeal of mountains and the specific engagement with sport climbing were two other phenomena, cartography and science. Many nations and colonial powers in Europe, Asia, Africa, and North and South America carefully mapped the rugged mountains, particularly the highest, uninhabited zones, which marked their boundaries and had not been measured precisely. Within science, research focused on glaciers (Orlove et al. 2008). This interest was prompted by natural disasters related to glaciers, such as the devastating floods of 1818 in the Rhone Valley in Switzerland, which resulted from the outburst of a lake that had been dammed by glacial ice. During the nineteenth century Swiss and Norwegian researchers began mapping glaciers. They drew on these observations and from the knowledge of local residents to propose the existence of an Ice Age in the remote past. Systematic data collection began in Switzerland in 1893 and in Norway a few years later. Swiss delegates to the Sixth International Geological Congress in 1894 established an International Glacier Commission, whose goal was to study long-term cycles of ice ages and shorter-term fluctuations of individual glaciers. Several glacier-related disasters in the Alps between 1892 and 1901 stimulated this interest. Other centers of glaciological research developed, as Sörlin (2009) shows for Sweden in the period 1930–55.

Deserts constitute our fourth and final case. Though Europe contains a number of arid areas, the true deserts, in the minds of Europeans, lay across the Mediterranean, south of the moist strip along the North Africa coast and associated highlands. Following the Arab–Muslim conquest of North Africa in the seventh and eighth centuries, trans-Sahara trade increased: gold, salt, and slaves traded north, and horses and manufactured goods shipped south. In subsequent centuries a number of Islamic kingdoms, such as Mali and Songhai, were based in the Sahel, just to the south of the Sahara. Knowledge of these kingdoms reached Europe, particularly through Leo Africanus, a Muslim born in Spain late in the fifteenth century who traveled extensively in North Africa and the Sahel; he was captured by Spaniards and brought to Rome, where he wrote geographical treatises and other works. His reports of Timbuktu, a large Muslim city in the Sahara, excited the imagination of Europeans.

The European exploration of the coasts of Africa began in the late fifteenth century and proceeded rapidly. Their movement into the interior occurred later, and the Sahara included some of the last lands to be reached by them. A group of British founded the Association for Promoting the Discovery of the Interior Parts of Africa in 1788. A number of the members had ties to abolitionism, to scientific exploration in the Pacific and elsewhere, and to commerce. They were eager to discover the source of the Niger and to locate Timbuktu, but the first confirmed visit to the city by a European did not occur until 1826. Trans-Saharan expeditions brought back materials for natural history and ethnographic museums. The Berlin Conference of 1884–85, also a kind of intergovernmental organization, marked the full partition of Africa among European powers, principally Britain and France; they controlled large desert areas, though their rule was centered in moister, more populated regions. They expressed concern that the Sahara might encroach on these regions, an idea later encapsulated in the notion of desertification, first described by Aubreville (1949). Though current scholarship views this concern as a misinterpretation of the local landscape largely disconnected from observation (Batterbury and Warren 2001; Fairhead and Leach 1996; Thomas and Middleton 1994), the image of the inexorable advancement of sand is pervasive in discussions of semiarid environments.

Much as in the other cases, deserts have had a long importance in world culture. They were seen as difficult, dangerous places, sites of adventure and daring, and often inhabited by ferocious tribal warriors. The popular film *Lawrence of Arabia* draws on earlier images (Caton 1999), as do frequent cartoons that depict one or more travelers crawling beneath a blazing sun across sandy wastes toward an oasis, usually a mirage (fig. 2.4). The nineteenth century was also a period of explorations of other deserts in the world, for example, in Central Asia, Australia, and elsewhere. Some deserts had economic importance as well: the Atacama Desert in South America contained large deposits of sodium nitrate, of great value as a fertilizer.

Period 2: Postwar Politics and Environmentalism

By the mid-twentieth century almost all of the four regions had changed from a colonial status and been incorporated into nation-states: as part of developed countries in the case of the Arctic and as a number of new small states for the low-lying islands. The conditions were more diverse for the mountains, existing largely as peripheral regions of countries in Europe, North America, and Asia, and for deserts, with cases including new states in the Sahara and Sahel and portions of other countries in several continents. This period saw the development of new cultural, intellectual, and political

2.4: Desert cartoon, with a crawling man and an oasis by Dave Carpenter. www.CartoonStock.com.

forces as well. The decades immediately after World War II were characterized by the Cold War and by the emergence of new nations, many of which were nonaligned and nearly all of which were members of the United Nations. The idea of development provided a framework within which these diverse nations could locate themselves and conduct their negotiations. Development also served well to orient the domestic policies of these nations, allowing citizens and nations alike to make claims for better material standards of living and for some elements of justice and democracy. Directly relevant to our cases is the ability of these development discourses to consider environmental issues. This period was further marked by a large expansion of scientific research, including the geosciences and ecology. The space race, promoted by the Cold War, led to an expansion of satellites, which provided data that led to an enormous increase in the understanding of the planet's physical and biological systems—much as the program of nuclear submarines during the Cold War promoted an understanding of the deep oceans and the sea floor.

Under the broad umbrella of development and of international institutions and supported as well by the expansion of environmental sciences, the four study regions were all engaged by the environmental discourses that began to emerge in the 1960s and 1970s. These engagements, in turn, shaped the later encounter of the four regions with climate change. We note that the ideas of climate change and sustainable development are both relatively old, the former dating to the late nineteenth century, when scientists traced the links among atmospheric carbon dioxide concentrations, radiation, and global temperatures (Weart 2003), and the latter to the late eighteenth and early nineteenth centuries, when developments in the study of soils and agriculture promoted attention to the cycling of nutrients and to the possibility of permanent soil degradation (Warde 2011). Supported by global environmental movements in the 1970s, climate change and sustainable development both achieved great prominence as explanatory and policy frameworks in the late 1980s, the former with a focus on greenhouse gases and their consequences, the latter with a broader scope of environmental and social systems.

The Stockholm Conference of 1972 was the first major environmental conference held by the United Nations, building on broad environmental social movements worldwide and the spirit of transformation and renewal of the 1960s. It was associated with the establishment of the United Nations Environmental Programme (UNEP) in the same year. It was followed twenty years later by the Rio Summit of 1992, also known as the Earth Summit and as the United Nations Conference on Environment and Development. These two summits have been crucial in shaping international debates about environmental issues, including sustainable development and climate change. They overlap with major climate organizations. The first is the Intergovernmental Panel on Climate Change (IPCC), jointly organized in 1988 by UNEP and the World Meteorological Organization. The three working groups of the IPCC, consisting of nationally nominated scientists and approved by the panel member countries, review scientific studies of climate and produce major reports every five years or so (see O'Reilly, this volume). These discuss observational data of changes that have already occurred, projections of anticipated future changes, the projected impacts of such changes on people and the environment, and analyses of adaptation and mitigation policies, with a strong focus on the twenty-first century. The reports represent scientific consensus and are fundamental references for climate research and policy development by international organizations and nation-states. We follow the convention of referring to them by their acronyms: FAR for the First Assessment Report (1990), SAR for the second (1995), TAR for the third (2001), AR4 for the fourth (2007), and AR5 for the fifth (2013–14). The UN established a convention on climate at the Rio Summit in 1992, together with conventions to protect biodiversity and to combat desertification. The climate change convention is known as the United Nations Framework Convention on Climate Change (UNFCCC). It holds annual Conferences of Parties, known by their acronym COP, where major issues are discussed, many NGOs and international organizations hold side events, and media coverage is extensive. The UNFCCC gave rise to the Kyoto Protocol, established in 1997. It was hoped that the Kyoto Protocol would lead to effective action in two areas: mitigation and adaptation. Initially scheduled to expire in 2012, it was extended till 2020. A legally binding agreement to include all UNFCCC nations is scheduled to be prepared by 2015 and to be enacted in 2020.

The 1972 Stockholm and 1992 Rio Summits also overlap with sustainable development as a framework for guiding policy and collective action. The term emerged from a UN commission that operated from 1983 to 1987, resulting in the report *Our Common Future*, generally known as the Brundtland Report (1987). The report defined sustainable development as "development that meets the needs of the present without compromising the ability of future generations to meet their own needs" and emphasized economic growth, social equity, and environmental protection. Its great generality gives it wide scope, covering water, energy, biodiversity, agriculture, and other systems, but also can diffuse its focus and make it difficult to assess and implement. The concept of sustainable development was highlighted in *Agenda 21*, the key document from the Rio Summit, and provided the central theme of the third summit, the United Nations Conference on Sustainable Development of 2012.

These organizations and programs have had strong effects in each of the four regions. In the case of the Arctic, by the time global attention to climate change was raised in the 1980s, the region was long established as a coherent region in the far north. The most widely recognized of the high-Arctic indigenous peoples, the Inuit, also known as Eskimos, are featured in over 150 patents and trademarks in the United States alone.[3] A new, crucial element of this global attention can be located in the mobilization of indigenous peoples in recent decades, focusing on environmental and political issues in the years before the emergence of global warming. One of the triggering events of this mobilization was the first Arctic Peoples Conference, organized in Copenhagen in 1973. It drew Inuit from Greenland and Canada, Sámi from Sapmi (Norway, Finland, and Sweden), and a diverse group of First Nations from Canada. The first Inuit Circumpolar Conference, held in 1977 in Barrow, Alaska, reflected the strong engagement of Alaskan Inuit as well and the growing influence of global indigenous movements. The conference issued seventeen resolutions focusing on autonomy, self-determination, and freedom of movement; several of them addressed environmental issues, particularly about hunting rights and wildlife conservation. Highly conscious of a major nuclear accident at Thule, Greenland, in 1968, the conference participants proposed banning nuclear weapons and waste from the Arctic. The awareness of transboundary movements of pollutants grew in 1986, when fallout from the explosion of the Chernobyl nuclear reactor reached the Arctic; radioactive substances accumulated in lichens, a principal food of reindeer, and affected the traditional Sámi diet. The Inuit were concerned about the high concentrations of persistent organic pollutants in the atmosphere in the 1990s and lobbied to reduce these threats to their well-being. The Inuit Circumpolar Conference issued declarations about persistent organic pollutants at their meeting in 1998, contributing to the efforts that led to the regulation of these pollutants by the Stockholm Convention of 2001.

Inuit communities have reported climate change, particularly in terms of temperature, ice conditions, and wildlife, since the 1980s and 1990s, and the Inuit Circumpolar Conference focused on the issue by the late 1990s. The organization, which renamed itself the Inuit Circumpolar Council in 2006, continues to be active in climate change issues and has sent delegates to a number of COPs, including COP 11 (Montreal, 2003), COP 13 (Bali, 2007), COP 15 (Copenhagen, 2009), COP 17 (Durban, 2011), and COP 18 (Doha, 2012), influencing, though not directly authoring, the official documents issued at these events. This involvement makes the Arctic one of the clearest instances of the active role of relatively disenfranchised groups in climate policy frameworks and institutions. The publication of cartoons that present climate change in the Arctic attests to the widespread perception of this region as being especially vulnerable (fig. 2.5).

Indigenous peoples have been active in the Arctic Council, with status as permanent participants, ever since its establishment in 1996. This consensus-based intergovernmental organization, composed of the eight high-Arctic countries, addresses issues of climate change as well as pollution, conservation,

2.5: Climate change cartoon in the Arctic, with Eskimo, igloo, and signs of warming by Felipe Galindo (Feggo). www.CartoonStock.com.

and sustainable development more broadly. It was instrumental in the Arctic Climate Impact Assessment of 2005, a major assessment of the impacts of climate change on Arctic environments and populations, with particular attention to the effects of climate change on indigenous peoples (ACIA 2005). The Inuit Circumpolar Conference, the Sámi Council, and the Russian Organization of Indigenous Peoples were active participants in preparing this assessment. The Arctic Council has not been very active in the COPs, in part because some member nations have resisted such efforts.

The international scientific community has long recognized that the Arctic is particularly sensitive to climate change because the rate and magnitude of change are greater there than in the rest of the globe. The Arctic has featured prominently in IPCC reports since the FAR and SAR of the 1990s, which often conflated the Arctic with Antarctica. The polar regions were accorded full chapters in TAR, AR4, and AR5. The reports paid particular attention to increasing temperatures and diminishing sea ice, to the thawing permafrost and to vegetation change. They also discussed impacts on human populations, including indigenous peoples, and reported easier transportation and increased income from petroleum production as positive features. TAR, AR4, and AR5 offered longer discussions of polar regions as a separate topic than earlier reports, broadening the scope to include human impacts and adaptation.

As in the Arctic, the low-lying islands have been involved in international environmental politics since the second half of the twentieth century, before the emergence of climate change as a global issue. Atmospheric tests of nuclear weapons on Pacific atolls began in the late 1940s, accompanied by the removal of some native populations and the exposure of others to radioactive fallout. In part owing to pressure from island groups, then still largely under colonial rule, a partial nuclear test ban treaty ended aboveground tests in 1963, though France continued them until 1996. Another vital environmental issue for island nations is fisheries management. The delimitation of territorial waters and of exclusive economic zones in oceans was expanded in the 1970s, initially as a result of pressure from Ecuador and Peru, which sought to gain control over their highly productive coastal fisheries. The Pacific Islands Forum Fisheries Agency, established in 1979 with the support of the South Pacific Forum, brought over a dozen island nations together to address regional fisheries issues (Barnett and Campbell 2010). In 1982 the Third United Nations Convention on the Law of the Sea extended the exclusive economic zone to two hundred nautical miles from coastlines, a move which granted large fishing zones to island nations and increased their bargaining power in international arenas.

In the 1990s two international entities, loosely building on the format and success of the Forum Fisheries Agency, were formed to represent island nations worldwide and specifically address climate change and sea level rise; they now work in close coordination. The first, the Alliance of Small Island States, or AOSIS, grew out of a conference on sea level rise organized by the government of Maldives, soon after major flooding there in 1987. The conference participants, including Kiribati, Trinidad and Tobago, Mauritius, and Malta, issued a declaration on global warming and sea level rise. Representatives from island nations attended the Second Climate Conference in Geneva in 1990, where the first IPCC Assessment Report was presented, and steps were taken toward the establishment of the UNFCCC in 1992. At this meeting, AOSIS was formed, with strong participation from Pacific and Caribbean participants (Heileman 1993). AOSIS members pressed for strong wording in the declarations issued by the Second Climate Conference, and the alliance has participated in the UNFCCC process to limit greenhouse gas emissions and to support adaptation for island nations (Roddick 1997). AOSIS currently has thirty-nine members, including island nations in the Pacific, the Indian and Atlantic Oceans, the Caribbean Sea, and a few Caribbean mainland countries (Belize, Guyana, and Suriname), as well as one African coastal country, Guinea-Bissau. It plays a very active role at the COPs and contributed to formulating the National Action Programmes for Adaptation, promulgated at COP 7 in Marrakech in 2001 (Barnett and Dessai 2002). Between 2004 and 2012 AOSIS contributed forty-four major documents to COPs and to the associated Subsidiary Bodies for Scientific and Technological Advice and for Implementation. AOSIS often evokes the powerful image of entire nations disappearing under rising seas, an event unprecedented in history and reminiscent of the myth of Atlantis (Lazrus 2012). As is the case for the Arctic, desert islands are presented in cartoons as sites where climate change has already advanced (fig. 2.6).

2.6: Climate change cartoon on a desert island, with a castaway, palm tree, and signs of sea level rise by Randy McIlwaine. www.CartoonStock.com.

AOSIS contributed to the second international entity, Small Island Developing States, or SIDS. Unlike AOSIS, which is an organization, SIDS is a category, officially recognized by the United Nations. AOSIS pressed for the inclusion of small island issues at the Rio Summit in 1992, which in turn gave rise to the Global Conference on the Sustainable Development of Small Island Developing States, held in Barbados in 1994. The Declaration of Barbados, issued at this conference, recognized the category of SIDS and considered a wide range of issues, including climate change, sea level rise, natural hazards, fisheries, and coastal management. Nearly all of the thirty-nine members of AOSIS are included in the fifty-one SIDS states, and the two groups work closely together.

The IPCC Assessments Reports have directed considerable attention to low-lying islands and to coastal regions, reflecting the strong engagement of relatively weak groups (in this case, small, poor nations) in influencing climate policy frameworks and institutions. These islands were discussed extensively in the FAR in a chapter on world oceans and coastal zones, much of the discussion centering on measuring and projecting sea level rise. The SAR contained a chapter on coastal zones and small islands. It mentioned the specific vulnerabilities of small island states stemming from physical, economic, and institutional factors and referred specifically to AOSIS. The TAR, AR4, and AR5 also each had a chapter dedicated to small islands. They discussed adaptation in detail, with express attention to the specific vulnerabilities of small island states and the potential for local institutions to support adaptation.

Mountains returned as a focus of global attention in the years after World War II. A Swiss geographer established a new glacier commission, the Permanent Service on the Fluctuations of Glaciers, within UNESCO in 1967; this organization planned a World Glacier Inventory in the 1970s, which began the World Glacier Monitoring Service in 1986. Its activities were supported by several contemporary developments, including more complete understanding of ice dynamics, satellite observations, and expanded computer networks. The public interest in the topic increased greatly in 1991, when hikers in the Alps discovered a man's body at the edge of an ice field. Research showed that his body, covered soon after his death with snow that turned into ice, had been preserved for over five thousand years until the glacier receded and exposed him.

Recent scientific and development programs in mountain regions stem from a program known as Man and the Biosphere (MAB). Proposed at a

UNESCO conference in Paris in 1968, MAB was launched in 1970 and played a major role in 1972 at the Stockholm Conference (di Castri 1976). In 1976 MAB began establishing a series of biosphere reserves, areas which are designated for protection in order to promote harmony between people and nature. Reflecting the strong influence of researchers from Alpine nations in Europe, MAB identified mountains as a priority area, and many MAB reserves were located in mountain regions in the Alps and elsewhere (Batisse 1993). The program sought to encourage economic development that maintained environmental values and upheld traditional cultural forms. A program to develop climate change research, centered in MAB mountain biosphere reserves, was established at a conference in Scotland in 2005 (Greenwood et al. 2005). These researchers concentrate on vegetated regions of mountains, located at lower elevations than glaciers. UNESCO-affiliated researchers were also instrumental in founding the Swiss-based International Mountain Society in the early 1970s. Backed partly by German development funds, the society publishes a journal on scientific and programmatic aspects of mountain environments. A similar set of concerns about preserving mountain environments from overexploitation was expressed by the International Mountaineering and Climbing Federation in its Kathmandu Declaration on Mountain Activities of 1982.

The International Centre for Integrated Mountain Development (ICIMOD) was founded in 1983 with support from UNESCO and Swiss and German aid programs (Orlove 2010). Based in Nepal, it centered on the Himalayas and the neighboring Hindu Kush. It promoted watershed management, off-farm income generation, and environmentally sensitive engineering, with the goal of avoiding erosion and other degradation of mountain environments. In later years it has addressed hazard risk reduction and the conservation of biodiversity as well. It remains active and fairly well funded, unlike the smaller, less stable mountain organizations in the Andes, East Africa, and elsewhere. The MAB mountain program pressed for the inclusion of mountain concerns at the Rio Summit. The conference report, *Agenda 21*, contains a chapter titled "Managing Fragile Ecosystems: Sustainable Mountain Development," which argues that mountain environments are being degraded—often irreversibly—through deforestation and overexploitation of agricultural and grazing land and links these causes to poverty and poor management. It emphasizes soil erosion as the most pressing problem, though it includes others, such as loss of biodiversity and deterioration of watersheds. It seeks to redress these problems through scientific research, the encouragement of environment-friendly technology, training programs, promotion of off-farm income sources, and participatory land use and watershed management. It proposes promoting these activities through the support of UN agencies, bilateral aid, the International Mountain Society, and regional organizations such as ICIMOD. These concerns and priorities match the framework of sustainable development as proposed in the Brundtland Report of 1987.

Though mountains were proclaimed a priority at the Rio Summit, no major mountain organizations were formed.[4] Instead, mountain issues are dispersed across a number of UN agencies. The International Year of Mountains was declared for 2002, with a major forum in Bishkek, Kyrgyzstan. The Mountain Partnership was set up to coordinate activities afterward, but it has accomplished relatively little. It held conferences in Italy in 2003 and Peru in 2004, but none after that. It serves as a loose umbrella group for bilateral and international organizations that work in mountainous areas, among which the Swiss have been particularly active. It publishes a newsletter, which since 2005 has appeared less and less frequently.

ICIMOD has begun to make mention of climate change and has attended most of the COPs since 2004, but its emphasis remains on land use and economic activity. It has not contributed documents to the COPs and their Subsidiary Bodies. In 2011 it joined the newly formed International Cryosphere Climate Initiative, a nascent NGO which seeks to address issues that link the Arctic with Antarctica and mountain glaciers and to take part in international climate negotiations. Mountains are mentioned in all the IPCC Assessment Reports, with attention centered on glacier retreat, well documented in recent decades and projected, with strong confidence, to continue. The reports point to the impacts of this retreat on the availability of water for agriculture and hydropower in regions adjacent to mountains. The SAR contained a chapter on mountains and suggested that agriculture and forestry were vulnerable. The TAR, AR4, and AR5 continue the emphasis on glacier retreat and hydrology. They introduce the issue of conservation, indicating that the ranges of plant and animal species have shifted upslope and will continue to do so, with the higher altitude species facing risks of extinction.

The emergence of global attention to deserts in the postwar period was also slow. Despite the scientific, cultural, and economic interest in deserts, no organizations proposing deserts as a distinct object of scientific inquiry were founded. Two major international centers, part of the Consultative Group on International Agricultural Research (CGIAR), address crop breeding and effective management of soils and water in semiarid areas: the International Center for Agricultural Research in the Dry Areas and the International Crops Research Institute for the Semi-Arid-Tropics, both founded in 1972. The former, which emphasizes the Middle East and North Africa, focuses on legumes, while the latter concentrates on South Asia and Sub-Saharan Africa and pays more attention to dryland grains, particular sorghum and millet.

Of great consequence to the place of deserts in climate change politics was the long African drought from the late 1960s through the early 1980s, beginning roughly a decade after many countries in the continent reached independence. It was most severe in the West African Sahel from 1968 to 1974 and all across the Sahel, including parts of Ethiopia in 1984. It served to reawaken earlier colonial worry about the spread of deserts, understood to be an irreversible march of sand into vegetated areas. The high mortality of people and livestock during this period created one of the great humanitarian crises of the postwar period.

Several organizations grew out of this period. One is a regional organization, the Permanent Inter-state Committee of Struggle against Drought in the Sahel, or CILSS. Its original nine members, whose territories are contiguous, consist largely of Francophone countries (Mauritania, Senegal, Burkina Faso, Mali, Niger, and Chad), though it also includes the Gambia, an Anglophone country, and two Portuguese-speaking countries (Guinea-Bissau and Cape Verde, which are both also members of AOSIS). It works on food security, water management, family planning, and other related issues. It receives support from overseas development assistance programs in the United States, Canada, and a number of European countries and from the OECD and coordinates with UN agencies as well. CILSS has recently begun to make some references to climate change, but this topic is a minor theme in its publications and activities, and it has attended only one COP in the last seven years.

The other organizations are associated with the United Nations. Faced with the drought in the Sahel, UNEP convened the UN Conference on Desertification in 1977 to conduct studies on drought and desertification and to develop plans to improve conditions in areas that had already suffered desertification. This conference proposed the UN Plan of Action to Combat Desertification (PACD), which was passed by the UN General Assembly in the same year. Its twenty-eight recommendations, covering a wide range of activities, proposed the establishment of national-level organizations to combat desertification, the running of workshops and training programs to address the problem, and the establishment of projects to halt desertification and promote recovery in affected areas. It shared the belief, widely held at the time, that the droughts in the Sahel and elsewhere were due to poor land-use practices; in the Sahel, these included overgrazing and shortening of fallow cycles. By the late 1980s PACD was widely faulted by UNEP and others for lack of coordination between the national and international levels, weak monitoring, poor financial planning, and technical failures. Others challenged the plan's weak scientific basis and vague, unworkable definitions of desertification (Rhodes 1991; Stringer 2008). PACD officials argued that their problems stemmed from lack of funding (Stiles 1984).

During the buildup to the highly visible Rio Summit, UNEP recognized the poor performance of PACD. The issue of desertification was raised repeatedly in Rio. A key chapter of *Agenda 21* was titled "Managing Fragile Ecosystems: Combating Desertification and Drought," repeating the title of the mountain chapter but substituting a different subtitle. This report also placed soils as a central concern, highlighting loss of fertility and degradation of soil structure rather than erosion. It attributed these changes to overuse originating in poverty and poor management, mentioning as well climate variability as a cause. It sought to redress these problems via scientific research, soil conservation, reforestation, training programs, promotion of alternative income sources, and participatory land-use management—a set of practices generally consistent with the framework of sustainable development, as enunciated in the Brundtland Report (1987). As a result, the UN created a second convention, the United Nations Convention to Combat Desertification (UNCCD). Though this convention sought to distinguish itself from its predecessor, above all by adopting the more current language of participation and decentralization, it too suffered from an imprecision in definition of deserts and desertification and from a lack of methods to monitor progress in reaching its goal of reducing or ending desertification (Rhodes 1991). The UNCCD had a membership of 193 countries. Some of these, such as Botswana, were largely desert, and others, such as Chile and Mongolia, had large desert regions. But others, not associated with desertification, used the convention to seek support. Moldova, claiming that crops there sometimes failed because of insufficient rains, requested funding, as did Guyana, a largely forested country with a small population, seeking to forestall any risk of land degradation.

The UNCCD encouraged the UN to declare 2006 the International Year of Deserts and Desertification, but this effort, too, did not achieve success. Other critiques of the UNCCD arose: the discussion of desertification led government officials to dwell on long-term changes in vegetation cover rather than on short-term drought, a greater concern to farmers (Slegers and Stroosnijder 2008). Moreover, climatological studies questioned the emphasis the UNCCD placed on land management by documenting the roots of the drought of 1968–74 in physical causes outside the region, particularly sea surface temperatures in the Atlantic and Indian Oceans (Giannini et al. 2003). However, the UNCCD continues to focus on desertification very broadly and to propose strategies that stress land use and water management to address the problem.

The scientific literature that links climate change and deserts is somewhat more tentative than that for the other regions discussed in this chapter, in part because of the hegemony of the human-induced desertification discourse that links drought primarily to mismanagement of natural resources, and in part because the research showing that climate change is already a reality in margins of the desert such as the Sahel is only a decade old at most. Greenhouse-gas induced warming of sea-surface temperatures can lead to drying in the Sahel and elsewhere (Giannini et al. 2003, 2008; Held et al. 2005), and other effects of climate change, particularly the increase in aerosols, may have exacerbated this effect in the 1970s and 1980s (Chang et al. 2011; Rotstayn and Lohmann 2002). Unlike the direct connections that link increasing greenhouse gas concentrations and rising temperatures with Arctic warming, sea level rise, and glacier retreat, desertification depends on rainfall decline as well as on warming, and the shifts in precipitation are not as well established or as well understood as the increases in temperature. Nonetheless, there is agreement that many (though not all) zones in the world are becoming drier and warmer and that these trends will continue (Liu et al. 2013; Sheffield et al. 2012).

The assessment reports have centered less directly on deserts than on polar regions and small islands, although they do discuss regions that experience or are projected to experience a decrease in precipitation, an increase in number and intensity of droughts, or both. By and large, drought and desertification are parallel discourses, developed, respectively, by Working Group I and II. Working Group I devoted attention to the Sahel in the FAR, while Working Group II contributed two chapters on deserts in the SAR, which traced the interactions of climatic and local human land-use practices in changing vegetation and soils. The TAR indicated areas that have increasing drought risk and drying; AR4 talked about these regions (in the subtropics, the Mediterranean, Central Asia, and elsewhere) in greater detail and expressed stronger confidence in its projections. It stressed the importance of human contributions in increasing vulnerability to impacts in these regions. Though TAR and AR4 indicate that a number of areas, including some semiarid zones bordering on deserts, are likely to become drier in coming decades, they describe these shifts as changes in precipitation or in droughts rather than as instances of desertification. This choice of terminology implies that climate change frameworks and desertification discourses remain separate.

This scientific basis seems sufficient to allow deserts to become strongly linked with climate change in the global public sphere. However, some factors have blocked the formation of this connection. Unlike the Arctic, low-lying islands, and glaciers, deserts are poorly defined, and the risks with deserts involve the spread of deserts to nondesert areas rather than the loss of deserts. More seriously, major international organizations maintain the notion that desertification is owing to local and regional human-induced land-use changes and that international aid, distributed nationally, can reduce the problem. Captured by other environmental and development discourses and organizations, deserts are not drawn as systematically as other places into climate change discussions. CILSS, the largest desert organization, maintains its ties to the United Nations through the UNCCD and has attended only two of the last eight COPs of the UNFCCC. Extensive searches of large cartoon databases have failed to turn up even a single cartoon linking mountains or deserts with climate change; whether these fruitless searches genuinely document an absence of such cartoons or merely indicate their great scarcity, they do hint that public attention has a spatial pattern similar to that of science and policy, stressing the importance of climate change in some regions and downplaying or excluding it in others.

The trajectories of these four regions, as presented in the documents of international organizations and more broadly in the global public sphere, fall into two categories.[5] The Arctic and low-lying islands have become closely associated with climate change. The changes these regions face are understood in the global public sphere, in science, and in intergovernmental organizations as the consequence of increased emissions of greenhouse gases. These cases are taken to demonstrate the urgency of climate change mitigation and the need for adaptation to climate change. Mountains and deserts, on the other hand, have been more closely linked to sustainable development, to the need for alternative land-use patterns and livelihoods, framed and funded differently from climate change adaptation. The mountain case is striking because it contains glaciers, also closely associated with climate change but less linked in public understandings and in international organizations to human populations.

We have traced these two different paths back to the long trajectories of engagement of all four regions with Europe and the West. These trajectories predominantly began with the onset of European expansion and were further shaped by the economic, political, scientific, and cultural dynamics of the colonial period. As a result of these histories, each region is understood as a highly specific kind of place, and each is associated with specific indigenous or localized populations. Though their political statuses varied in the postwar period (the islands and deserts as newly independent nations, the Arctic as a set of high-latitude areas of developed nations, and the mountains largely as peripheries of lowland states), all four engaged during the sixties, seventies, and eighties with postwar environmental discourses. These engagements set the course for their highly distinctive involvements with the sustainable development and climate change frameworks, which emerged fully in the late 1980s and early 1990s—a period which opened with the declaration of sustainable development in the Brundtland Report of 1987 and the foundation of the IPCC in 1988 and ran to the Rio Summit of 1992, which produced key documents and accords. The bifurcation which allocates specific regions to one framework or the other has continued in the twenty-first century.

The mountains were the earliest to begin these engagements, promoted by scientists at the UNESCO Man and the Biosphere program and at the Stockholm Summit of 1972. The deserts entered second, after the devastating Sahel droughts, via UNCOD in 1977. Both of these were discussed extensively in *Our Common Future*, the report issued by the Brundtland Commission. These two cases were featured prominently in the Rio Summit, whose conference report spoke of both as fragile ecosystems to be managed through sustainable local land-use practices; these practices would prevent desertification of semiarid areas and erosion of the thin soils on steep mountain slopes. These two cases shared a focus on soils and sought local remedies to the poverty that drove overexploitation of land (with a new UN convention in the case of deserts). Though this framing emphasized poverty alleviation and called for international organizations to support appropriate development projects, it also placed the responsibility for the environmental problems with the local inhabitants whose practices supposedly degraded local environments.

It proposed local development and improved environmental management as solutions. Desert organizations like CILSS and mountain organizations like ICIMOD have maintained this focus.

The Arctic and low-lying islands both had significant postwar histories of regional environmental movements with strong indigenous participation, concerned with nuclear contamination and persistent organic pollutants in the former and with nuclear testing and fishing rights in the latter. (The latter were mentioned briefly in the Brundtland Report, which makes no reference at all to the Arctic.) These movements placed responsibility outside the regions—with nuclear powers, with the industrial nations who produced pollution that reached the Arctic, with powerful nations whose governments and commercial fishing fleets exploited resources that properly belonged to island nations. They proposed new international policies as solutions. The framings were continued directly into climate issues. To be sure, these two cases differ in some details. The small island states joined early in the movements that gave rise to the UNFCCC and the Kyoto Protocol and have played a major role in them. The Inuit Circumpolar Council and the Arctic Council pressed the UNFCCC from outside. Compared to the great spatial dispersal of the low-lying islands, the Arctic forms a single region and has benefited from the geographical continuity, even though it is spread across the tops of eight countries, spanning three continents and several major islands. Nonetheless, the two groups have played a powerful role in international discussions. The recognition of this similarity has extended into the world of cartoons, as illustrated by a recent magazine cover that links the two cases (fig. 2.7).

These divergent outcomes reflect a combination of disparate factors. Some of these have deep historical roots, such as the nineteenth-century scientific organizations that studied the Arctic and glaciers and earlier cultural framings of islands as exotic alternative worlds, of deserts as hostile wastelands, of mountains and the Arctic as pristine and sublime. Some of these are more recent, such as postwar environmental movements and the availability of sustainable development framings and organizations, which influence international aid. Science, too, has played a role. In particular, the linkages of climate change to the Arctic, glaciers, and low-lying islands are quite direct through changes in temperature, which melt ice in the first two cases, contributing to sea level rise, which affects the third. The connections with desertification center on shifts in precipitation, some from the alteration of broad precipitation belts globally, some from the complex regional connections between ocean warming and storm patterns.[6] The influences between

2.7: Climate change cartoon, linking Arctic and desert island themes. Ian Falconer for *The New Yorker*. Courtesy of *The New Yorker*.

science and politics operate in both directions. Though the claims about the vulnerability of low-lying islands in the IPCC reports are well established empirically, the great attention to these islands derives in part from the strong influence of AOSIS, much as the lack of discussion of desertification in these reports reflects the weaker presence of CILSS, and from the existence of the UNCCD as an alternative to the UNFCCC.

Can any insight into the future of climate change be derived from the recognition that the discussions of climate change have long histories and involve many groups? In the current period of transition from the Kyoto Protocol to its as-yet uncreated successor, it is difficult to discern the outlines of the policies and institutions that will shape future responses to climate change. Nonetheless, some programs have begun to address the impacts in these specific regions; early examples include the planned relocation, building from around 2000 and increasing in activity around 2007, of coastal Native villages in Alaska, plagued by coastal erosion owing in part to sea ice loss (Yardley 2007). It is striking that in this area of long indigenous political activism on environmental issues, both the scientific community and policy makers have entered—more extensively than in other areas—into dialogue with indigenous peoples (for example, Bravo 2009; Laidler 2006). The long discussions over resettlement of populations from atolls in the Pacific affected by sea level rise to New Zealand, which achieved recognition in the Niue Declaration made at the Pacific Islands Forum in 2008, are also notable (McAdam 2012).

Our review of the four regions has shown the historical depth of the engagements that join them to colonial, international, and global orders. Though these engagements demonstrate the great unevenness of power in global economic and political systems, they also show the capacity of some small groups in remote areas to gain wide recognition, to pressure powerful nations to take responsibility for their actions, and to influence intergovernmental organizations that manage issues such as nuclear testing, fishing rights, and pollution, generally in the direction of greater equity and sustainability. In the earlier decades when these issues unfolded, the urgency and difficulty of establishing a new order seemed broadly as challenging as the tasks that climate change now raises—grounds, perhaps, for optimism. However, our review of these regions also shows the lasting power of nonclimate framings to limit recognition of climate impacts in other cases, such as the mountains and deserts, and to deflect responsibility for these regions. In this complex terrain of recognition and responsibility future climate politics will unfold.

NOTES

We would like to thank Kaitlin Butler, Nick Cox, and Julia Olsen for their work as research assistants and to express our gratitude to the many people who commented on earlier versions of this chapter presented at the University of Copenhagen, the Graduate Center of the City University of New York, and Yale University. Mattias Borg and Pam McElwee gave particularly detailed and valuable comments on these earlier versions. An article which draws on the four cases presented here, emphasizing the spatial patterning of the regions rather than the temporal sequence that provides the focus of this chapter, was published as Orlove et al. 2014.

1. Climate change is a complex process, distinct from natural variability, and it interacts with changes in environmental, economic, and societal conditions. Even the processes that appear most directly and exclusively tied to climate change are associated with other dynamics. Similarly, other environmental problems—for example, loss of

biodiversity and contamination of water and air—have causes separate from those of climate change, though they can be exacerbated by climate change.

2. Currently, a more commonly used term for this is the Northern Sea Route.
3. The registry of the United Stated Patent and Trademark Office lists 162 trademarks since 1905 and 2 patents since 1976 associated with the term *Eskimo*, including the chocolate-covered ice cream known as Eskimo Pie. The Australian Trade Marks Online Search System at IP Australia indicates that the term *Esky* for the portable cooler was trademarked in 1961, though similar coolers under the same name were first produced in 1884 (http://superbrands.com.au/index.php/volumes/volume-2/79-volumes/volume-2/247-esky-vol-2).
4. Geopolitics may have played a role in this absence. Though a few countries such as Nepal and Bhutan promote themselves as mountain nations, they do not have the number or weight of small island nations. The Swiss support mountain development internationally but cannot construct a group as powerful as the Arctic Council because some of the most important potential members, like China, India, and Pakistan, are more preoccupied with international tensions in the Himalayas.
5. In a realist sense, all four regions are affected both by climate change and by the problems associated with unsustainable development. However, we seek to emphasize, in a constructivist sense, that regions are tied discursively and organizationally to one or the other of the frameworks.
6. The Arctic and low-lying islands also face other serious environmental issues that are loosely connected to climate change, or quite distant from it, such as the expansion of petroleum drilling and solid waste management. These examples indicate that sustainable development frameworks are applicable to these regions, much as climate change frameworks are applicable to mountains and deserts.

REFERENCES CITED

ACIA. 2005. *Arctic Climate Impact Assessment.* Cambridge: Cambridge University Press,

Aubreville, A. 1949. *Climats, Forêts et Désertification de l'Afrique Tropicale.* Paris: Société d'Editions Géographiques, Maritimes et Coloniales.

Barnett, J., and J. Campbell. 2010. *Climate Change and Small Island States: Power, Knowledge and the South Pacific.* London: Earthscan.

Barnett, J., and S. Dessai. 2002. "Articles 4.8 and 4.9 of the UNFCCC: Adverse Effects and the Impacts of Response Measures." *Climate Policy* 2, nos. 2–3: 231–39.

Batisse, M. 1993. "The Silver Jubilee of MAB and Its Revival." *Environmental Conservation* 20, no. 2: 107.

Batterbury, S., and A. Warren. 2001. "Desertification." In *International Encyclopædia of the Social and Behavioral Sciences*, ed. N. Smelser and P. Baltes, 3526–29. Oxford: Elsevier Press.

Beck, U. 1999. *World Risk Society.* Malden: Polity Press.

Bravo, M. T. 2009. "Voices from the Sea Ice: The Reception of Climate Impact Narratives." *Journal of Historical Geography* 35, no. 2: 256–78.

Castells, M. 2008. "The New Public Sphere: Global Civil Society, Communication Networks, and Global Governance." *Annals of the American Academy of Political and Social Science* 616: 78–93.

di Castri, F. 1976. "International, Interdisciplinary Research in Ecology: Some Problems of Organization and Execution. The Case of the Man and the Biosphere (MAB) Programme." *Human Ecology* 4, no. 3: 235–46.

Caton, S. C. 1999. *Lawrence of Arabia: A Film's Anthropology.* Berkeley: University of California Press.

Chang, C. -Y., J. C. H. Chiang, M. F. Wehner, A. R. Friedman, and R. Ruedy. 2011. "Sulfate Aerosol Control of Tropical Atlantic Climate over the 20th Century." *Journal of Climate* 24, no. 10: 2540–55.

Crutzen, P. J. 2006. "The 'Anthropocene.'" In *Earth System Science in the Anthropocene,* ed. E. Ehlers and T. Krafft, 13–18. New York: Springer-Verlag.

Cushman, G. T. 2013. *Guano and the Opening of the Pacific World: A Global Ecological History.* Cambridge: Cambridge University Press.

Demeritt, D. 2001. "The Construction of Global Warming and the Politics of Science." *Annals of the Association of American Geographers* 91, no. 2: 307–37.

Doulton, H., and K. Brown. 2009. "Ten Years to Prevent Catastrophe? Discourses of Climate Change and International Development in the UK Press." *Global Environmental Change* 19, no. 2: 191–202.

Ericson, J., C. Vorosmarty, S. L. Dingman, L. Ward, and M. Meybeck. 2006. "Effective Sea-Level Rise and Deltas: Causes of Change and Human Dimension Implications." *Global and Planetary Change* 50, nos. 1–2: 63–82.

Fairhead, J., and M. Leach. 1996. *Misreading the African Landscape: Society and Ecology in a Forest-Savanna Mosaic.* Cambridge: Cambridge University Press.

Farbotko, C., and H. Lazrus. 2012. "The First Climate Refugees? Contesting Global Narratives of Climate Change in Tuvalu." *Global Environmental Change* 22: 382–90.

Giannini, A., R. Saravanan, and P. Chang. 2003. "Oceanic Forcing of Sahel Rainfall on Interannual to Interdecadal Time Scales." *Science* 302: 1027–30.

Giannini, A., M. Biasutti, and M. M. Verstraete. 2008. "A Climate Model-Based Review of Drought in the Sahel: Desertification, the Re-Greening and Climate Change." *Global and Planetary Change* 64, nos. 3–4: 119–28.

Gillis, J. 2004. *Islands of the Mind: How the Human Imagination Created the Atlantic World.* New York: Palgrave Macmillan.

Greenwood, G., A. Björnsen, C. Drexler, and M. Price. 2005. "MRI Newsletter 5: GLOCHAMORE Update." *Mountain Research and Development* 25, no. 3: 282–83.

Grove, R. H. 1995. *Green Imperialism: Colonial Expansion, Tropical Island Edens and the Origins of Environmentalism, 1600–1860.* Cambridge: Cambridge University Press.

Gupta, A. 1998. *Postcolonial Developments: Agriculture in the Making of Modern India.* Durham: Duke University Press.

Hacking, I. 1999. *The Social Construction of What?* Cambridge: Harvard University Press.

Hardin, R. 2011. "Concessionary Politics: Property, Patronage and Political Rivalry in Central African Forest Management." *Current Anthropology* 52, no. S3: S113–S125.

Heileman, L. 1993. "The Alliance of Small Island States (AOSIS): A Mechanism for Coordinated Representation of Small Island States on Issues of Common Concern." *Ambio* 22, no. 1: 55–56.

Held, I. M., T. L. Delworth, J. Lu, K. L. Findell, and T. R. Knutson. 2005. "Simulation of Sahel Drought in the 20th and 21st Centuries." *Proceedings of the National Academy of Sciences* 102, no. 50: 17891–896.

Hovelsrud, G. K., B. Poppel, B. van Oort, and J. D. Reist. 2011. "Arctic Societies, Cultures, and Peoples in a Changing Cryosphere." *Ambio* 40, no. 1: 100–110.

Hughes, D. M. 2013. "From Oil to Innocence: Climate Change and the Victim Slot." *American Anthropologist* 115, no. 4: 570–81.

Jasanoff, S. 2010. "A New Climate for Society." *Theory, Culture and Society* 27, no. 2: 233–53.

Laidler, G. J. 2006. "Inuit and Scientific Perspectives on the Relationship Between Sea Ice and Climate Change: The Ideal Complement?" *Climatic Change* 7, nos. 2–4: 407–44.

Latour, B. 2004. "Why Has Critique Run Out of Steam? From Matters of Fact to Matters of Concern." *Critical Inquiry* 30, no. 2: 225–48.

Lazrus, H. 2012. "Sea Change: Island Communities and Climate Change." *Annual Review of Anthropology* 41: 285–301.

Leiserowitz, A. 2005. "American Risk Perceptions: Is Climate Change Dangerous?" *Risk Analysis* 25, no. 6: 1433–42.

Liu, J., B. Wang, M. A. Cane, S. -Y. Yim, and J. -Y. Lee. 2013. "Divergent Global Precipitation Changes Induced by Natural Versus Anthropogenic Forcing." *Nature* 493: 656–59.

Lorenzoni, I., A. Leiserowitz, M. de Franca Doria, W. Poortinga, and N. F. Pidgeon. 2006. "Cross-National Comparisons of Image Associations with 'Global Warming' and 'Climate Change' Among Laypeople in the United States of America and Great Britain." *Journal of Risk Research* 9, no. 3: 265–81.

McAdam, J. 2012. *Climate Change, Forced Migration, and International Law.* Oxford: Oxford University Press.

Mann, M. E. 2002. "Medieval Optimum Climate." In *The Earth System: Physical and Chemical Dimensions of Global Environmental Change*, ed. M. C. MacCracken and J. S. Perry, 1:514–16. Chichester: John Wiley and Sons.

Manzo, K. 2010. "Imaging Vulnerability: The Iconography of Climate Change" *Area* 42, no. 1: 96–107.

Mathews, A. 2011. *Instituting Nature: Authority, Expertise, and Power in Mexican Forests.* Cambridge: MIT Press.

Nicolson, M. H. 1959. *Mountain Gloom and Mountain Glory: The Development of the Aesthetics of the Infinite.* Ithaca: Cornell University Press.

Orlove, B. 2009. "The Past, the Present, and Some Possible Futures of Adaptation." In *Adaptation to Climate Change: Thresholds, Values, Governance*, ed. W. Neil Adger, Irene Lorenzoni, and Karen O'Brien, 131–63. Cambridge: Cambridge University Press.

———. 2010. "Glacier Retreat: Reviewing the Limits of Adaptation to Climate Change." *Environment* 51, no. 3: 22–34.

Orlove, B., W. Ellen, and B. H. Luckman, eds. 2008. *Darkening Peaks: Glacial Retreat, Science and Society.* Berkeley: University of California Press.

Orlove, B., H. Lazrus, G. K. Hovelsrud, and A. Giannini. 2014. "Recognitions and Responsibilities: On the Origins and Consequences of the Uneven Attention to Climate Change Around the World." *Current Anthropology* 55(3): 249–75.

Reynolds, T. W., A. Bostrom, D. Read, and M. G. Morgan. 2010. "Now What Do People Know About Global Climate Change? Survey Studies of Educated Laypeople." *Risk Analysis* 30, no. 10: 1520–38.

Rhodes, S. 1991. "Rethinking Desertification: What Do We Know and What Have We Learned?" *World Development* 19, no. 9: 1137–43.

Roddick, J. 1997. "Earth Summit North and South: Building a Safe House in the Winds of Change." *Global Environmental Change* 7, no. 2: 147–65.

Rotstayn, L. D., and U. Lohmann. 2002. "Tropical Rainfall Trends and the Indirect Aerosol Effect." *Journal of Climate* 15, no. 15: 2103–16.

Sheffield, J., E. F. Wood, and M. L. Roderick. 2012. "Little Change in Global Drought Over the Past 60 Years." *Nature* 491: 435–38.

Slegers, M. F. W., and L. Stroosnijder. 2008. "Beyond the Desertification Narrative: A Framework for Agricultural Drought in Semi-Arid East Africa." *Ambio* 37, no. 5: 372–80.

Smith, N., and H. Joffe. 2009. "Climate Change in the British Press: The Role of the Visual." *Journal of Risk Research* 12, no. 5: 647–63.

Sörlin, S. 2009. "Narratives and Counter-Narratives of Climate Change: North Atlantic Glaciology and Meteorology, c. 1930–1955." *Journal of Historical Geography* 35, no. 2: 237–55.

Stiles, D. 1984. "Desertification—The Time for Action." *The Environmentalist* 4, no. 2: 93–96.

Stringer, L. C. 2008. "Reviewing the International Year of Deserts and Desertification 2006: What Contribution Towards Combating Global Desertification and Implementing the United Nations Convention to Combat Desertification?" *Journal of Arid Environments* 72, no. 11: 2065–74.

Swyngedouw, E. 2010. "Apocalypse Forever? Post-Political Populism and the Spectre of Climate Change." *Theory, Culture, and Society* 27, nos. 2–3: 213–32.

Thomas, D. S. G., and N. Middleton. 1994. *Desertification: Exploding the Myth.* Chichester: Wiley.

Tsing, A. L. 2005. *Friction: An Ethnography of Global Connection.* Princeton: Princeton University Press.

Warde, P. 2011. "The Invention of Sustainability." *Modern Intellectual History* 81: 153–70.

Weart, S. R. 2003. *The Discovery of Global Warming.* Cambridge: Harvard University Press.

West, P. 2006. *Conservation Is Our Government Now: The Politics of Ecology in Papua New Guinea.* Durham: Duke University Press.

Wolf, E. R. 1982. *Europe and the People without History.* Berkeley: University of California Press.

Yardley, W. 2007. "Victim of Climate Change, a Town Seeks a Lifeline." *New York Times,* May 27.

Chapter 3 **From Conservation and Development to Climate Change: Anthropological Engagements with REDD+ in Vietnam**

Pamela McElwee

One of the most broadly discussed options on the table for the next phase of the Kyoto Protocol, which sets targets for global greenhouse gas levels for signatory countries, is a measure for Reduced Emissions from Deforestation and Degradation (REDD+) to tackle land-use-generated emissions.[1] Because forests serve as a sink for the absorption of around one-third of anthropogenic carbon emissions and have contributed around 8–10 percent of yearly anthropogenic carbon emissions when burned or cut, including them in target planning is a goal for many participants in the global climate convention (Gullison et al. 2007). The REDD+ program has been in active discussion since the Bali Conference of the Parties in 2007 agreed that the program would be a significant impetus to slow world deforestation. The fundamental premise of REDD+ is that if households and governments are given payments and other types of rewards that equal or exceed what could be raised by cutting down trees, then forests will be better protected and carbon conserved.

Yet as this global architecture for REDD+ develops, analysis is needed to determine how different REDD+ will be from the numerous forest protection policies that have come before. Anthropologists have long raised questions about the intersection between local people and forest conservation, including the ways in which power and culture often collide in forest conflicts between communities and the state (Sivaramakrishnan 2000; Haenn 2002); how communities living in forests are represented to global audiences and how they represent themselves vis-à-vis environmental problems (Haenn 1999; Brosius 2002; Doane 2007); and how such problems as deforestation come to be recognized as being in need of intervention in the first place (Dove 1983; Brosius 2006; Mathews 2008; McElwee, forthcoming). State policies for forest management have often been directed at restricting local forest use, resettling people away from degraded areas or instituting protected area boundaries, which have often had detrimental impacts on affected communities; anthropologists have long engaged in both research and activism to highlight these issues (Baviskar 2000; Laungaramsri 2002; West et al. 2006).

Similar anthropological attention is needed regarding the development of REDD+ and other climate interventions. To date, much of the emerging literature on REDD+ has focused on technical issues, including how forests will be defined and measured; how baselines against which progress in halting deforestation will be measured; how REDD+ financing will operate; and how benefits will be shared (Agrawal et al. 2011; Corbera and Brown 2010; Corbera and Schroeder 2011; Visseren-Hamakers et al. 2012). Rather than pursue these lines of inquiry, however, this chapter aims at two broader goals: one, to look more comprehensively at how REDD+ has grown out of a long line of interest in tropical forest management and, in so doing, to explore how new climate motivations may be similar to previous policies directed at tropical forests; and two, to use ethnographic tools to explore the development of REDD+ projects and assess what anthropology can contribute to understanding the implementation of REDD+ as a climate mitigation scheme (see also Mathews, this volume). I argue that major barriers to implementation of REDD+ are likely to revolve around questions regarding rights to forest environments and responsibilities of forest-using peoples. Previous forest conservation approaches have foundered on these questions, and REDD+ may face the same fate. Yet so far, REDD+ projects have ignored these deeper issues and focused primarily on seemingly apolitical technical requirements to identify and classify the myriad local conditions REDD+ projects might face, from land tenure to carbon content to social participation, a problem I term *checklist-ification.*

Using a case study from Vietnam, I examine the history of forest management and recent development of REDD+ and highlight similarities between the new climate-inspired policies and previous, usually unsuccessful forest policies, including Integrated Conservation and Development Projects and decentralized forest management. I ask what it means for our understanding of REDD+ in Vietnam if it is indeed the latest, principal medium of articulation of global, national, and local actors contesting how forests should be defined and who has rights to control them. The chapter points out that just as previous approaches often left the most vulnerable forest dwellers excluded from resources and benefits, new REDD+ policies have the potential to do the same, providing a cautionary tale for climate-driven development.

FROM TIMBER MANAGEMENT TO BIODIVERSITY CONSERVATION TO ECOSYSTEM SERVICES IN VIETNAM'S POLITICAL FORESTS

Southeast Asia is home to some of the richest forest resources in the world, and a history of forestry policies here shows that ideas about how best to manage forests come in waves, with remarkable similarities across the political spectrum of state governments in the region. As Peluso and Vandergeest have noted, colonial and then independent governments often undertook a series of similar actions regarding forests in the twentieth century: they all declared that the state was the sovereign owner of land and resources, enacted laws for demarcation and mapping of what constituted forests, and controlled the harvesting, trade and use of specific forest products, particularly valuable ones (Peluso and Vandergeest 2001). A consequence of these actions was the criminalization of previous forest management approaches, such as swidden agriculture, which was a particular enemy of nascent forest services, often based on inaccurate understandings of this system of land management (Dove 1983).

These new delimited areas—what Peluso and Vandergeest have termed "political forests"—were established in French Indochina in the early twentieth century, and by 1939 there were 2,250,000 hectares (ha) of reserved and protected forest (Service scientifique de l'Agence économique de l'Indochine 1931). As elsewhere, in Indochina the Forest Service made great efforts to try to keep ethnic minority peoples out of state-designated forest lands owing to fears about their use of swidden, although these restrictions were much contested by local peoples (McElwee, forthcoming). Despite the efforts of a few isolated colonial officials who argued for decentralized responsibility to local villages for forest management, most state approaches were characterized by hostility to local land practices and by attempts at centralized control of the forest estate (Thomas 2009).

Shortly after the Democratic Republic of Vietnam (DRV) was founded in the north in 1954, forest policy aimed at the complete nationalization of the forest estate and the establishment of State Forest Enterprises (SFEs) to log these lands. The nationalization of forests was extended to the south after 1975 at the conclusion of the Vietnam War, and there were more than four hundred SFEs at the height of state control of forests in the early 1980s (Ngo et al. 2006). The local peoples who had used and managed forest lands before nationalization received no financial remuneration for their land losses, and ethnic minorities in particular, who traditionally resided in much of the uplands where the richest forests were, were deeply affected. The combination of poor management in the SFEs combined with loss of official rights to forests among upland communities resulted in incentives for deforestation in many areas (To 2009). In the war years, from 1943 to 1976, Vietnam's forest cover declined from 14.3 million ha of forest to 11.1 million ha (Forest Inventory and Planning Institute 1996).

The cessation of war did not reduce the rapid deforestation; indeed, as Vietnam tried to rebuild and rejoin the world economy, deforestation increased considerably. Concern over illegal logging was voiced with increasing frequency in the Vietnamese press in the 1990s, focusing on the phenomenon of the *lam tac*, or illegal logger, someone who poaches and deforests with impunity, usually because the person has connections to political and economic elites (McElwee 2005; Sikor and To 2011). Landslides, flooding, and other environmental disasters throughout Southeast Asia in the 1990s focused further attention on watersheds and forests' perceived (though often exaggerated) role in conservation and climate regulation (Forsyth and Walker 2008). These concerns resulted in subsequent bans on timber imports and exports in many neighboring nations, and Vietnam too banned not only raw log exports but also many areas from commercial logging in 1993 (Tuynh and Phuong 2001). Yet the logging ban only regarded one aspect of the problem, ignoring other culprits of deforestation: the uneven land tenure situation among households living near forests; little support for communal management of forests, above all by ethnic minorities; incentives for illegal logging given high domestic demand for timber; mass migration of Vietnamese to upland areas, where they converted forests to agriculture; and poor management and enforcement of lands under continued SFE control (De Koninck 1999;

McElwee 2005; Meyfroidt and Lambin 2008; Meyfroidt and Lambin 2009; McElwee, forthcoming).

Coupled with the intensified interest shown in the region's forests by global NGOs concerned about wildlife and species preservation, the 1990s saw many forest ministries reinventing themselves as champions of biodiversity in areas they may have once overexploited. New protected areas that were to be free from exploitation grew substantially across the region (Déry and Vanhooren 2011). Yet despite appearances that these efforts to create strict protected areas were based on science alone, they tended to be concentrated in forested mountain areas inhabited by ethnic minorities, and certain groups, such as the Hmong in Thailand and Vietnam, were singled out and vilified as "forest destroyers" (Laungaramsri 2002; Forsyth and Walker 2008). In Vietnam, swiddeners continued to be targets of government sedentarization programs to substitute wet rice and cash crops for swidden fields.

More inclusive models of resource management, which were supposed to link conservation to poverty reduction, began to arise in many parts of the globe, usually under the name of "integrated conservation and development projects" (ICDPs) (Brandon and Wells 1992). In Vietnam, ICDPs were mainly funded by foreign donors, while the Vietnamese state kept to a strict policy of no human use of core areas of protected zones (although in reality many parks existed merely on paper) (McElwee 2002; Rugendyke and Son 2005). Examples of some of the emergent ICDP arrangements that donors brought to Vietnam included establishment of mixed-use "buffer zones" around core protected areas (Gilmour and Nguyen 1999); funding for low-impact, nontimber forest product extraction (Raintree 2004); and local comanagement of protected areas (Spelchan et al. 2011). However, the ICDPs had mixed results. Projects were often not scaled to the threats the forests actually faced, such as global demand for endangered timber and wildlife. Linkages with development were also weak, resulting in resentment and poor participation, especially when the benefits of ICDPs were not equally shared among community members (McElwee 2010; Polet 2003; Zingerli 2005). Many of Vietnam's ICDPs were eventually declared failures.

The ICDP approaches, although many were less successful than had been hoped, did coincide with increased enthusiasm for decentralization in forest management, with communities and households taking over the management of forests and other conservation areas from previous state or private landowners. Many Southeast Asian nations experimented with different models of decentralization and community and "social" forestry (Colfer

2012; Dove 1995). Vietnam too began to move in this direction, particularly with regard to forests not included in protected areas. Starting in 1993 with revisions to the national Land Law, citizens could receive fifty-year land certificates to forest land and the rights to manage the land in accordance with the land classification it was under, which usually meant they agreed to replant forestry land with trees or let degraded forest recover (McElwee 2009). The forest decentralization program provided land tenure certificates, commonly known as a "red book," and financial assistance to help smallholders reforest their lands for production forestry, taking land primarily from SFEs. However, major questions were raised about what institutional arrangements should be used to manage forest lands: allocation to individual households or to groups of households and communities, which led to decentralization of forest lands that was often uneven and incomplete (Clement and Amezaga 2009). As of the Rural Agricultural Census of 2006, 92 percent of the nation's agricultural land was held by private households, while only 24 percent of officially classified forest land was held by nonstate entities (households or communities) with land tenure rights (GSO 2007). Household rights to forests also remain unevenly distributed regionally. In addition, the forest lands which were decentralized were almost without question the poorest quality forest lands, as most land tenure certifications given out were for lands already denuded of any tree cover by SFEs (Thuan 2006). There have also been mixed results with the individual land rights that have been allocated. Such allocations have inevitably been embedded in local power and economic relations that often disadvantaged the poorest and those least able to interact with the state apparatus for allocation, namely, ethnic minorities (Sowerwine 2004; Sikor and Tran 2007; McElwee 2009; Coe 2012).

Given the frustrating lack of clear successes with decentralization, a new forest policy approach began in the mid-2000s: market-based incentives for conservation, like payments for environmental services (PES), which were designed to arrest degrading forest processes by providing economic valuation of important ecosystem functions (Wunder et al. 2008). Vietnam began to experiment with PES for landscape protection, ecotourism user fees, and carbon sequestration, and by 2012 at least nineteen PES projects were operating within Vietnam, sponsored by large donors like the World Bank and United Nations Development Program as well as conservation organizations like the World Wildlife Fund (Pham et al. 2009; McElwee 2012; To et al. 2012). Two state-run PES pilot projects were set up in Lam Dong and Son La provinces in 2008, and in 2010 the state extended PES policy nationwide through Decree 99, which aims to collect payments from hydropower companies, water companies, industrial facilities that use water, and tourist companies (Pham et al. 2013; McElwee et al. 2014). Forest service suppliers have included households and individuals near forest lands who have either protection contracts (essentially labor contracts agreeing to watch over state lands in return for cash) or red books (secure land tenure for forests), as well as state-owned forest entities, such as SFEs. Significantly, the two pilots to test PES were located in areas where ethnic minorities are the main forest users or owners (Thai/Dao/Hmong groups in Son La and Koho/Chil/Mnong in Lam Dong). While PES project documents have noted that this is an attempt to target the poorest households, the long history of the state's dislike of forest use by minorities suggests that these areas were targeted for reasons of identity as much as for forest threats. It remains to be seen if these market-oriented PES projects will be able to improve forest conservation outcomes, as current research on the policy indicates continued problems with uneven land tenure and benefit distribution (McElwee 2012; To et al. 2012; Pham et al. 2013).

THE DEVELOPMENT OF REDD+ AS EMERGENT FOREST POLICY

The rise of market mechanisms to protect forests has in more recent years coincided with concerns about slowing or halting land-use emissions as part of climate mitigation efforts. As Michael Dove (2003) has noted, the idea that tropical forests are the lungs of the planet is a long-standing one. However, this concept has taken a long time to develop into global climate policy. Early discussions regarding the Clean Development Mechanism of the Kyoto Protocol, which allows developed countries to transfer funding and technology to developing countries and claim emissions reductions realized in those countries as part of their own national emissions reductions, centered on strong political arguments against including forestry activities for fear that this would lower the quality of carbon credits (Fry 2008). Eventually reforestation and afforestation projects were made eligible for Clean Development Mechanism funding and emissions reductions credits (although they comprise but a small portion of the overall portfolio), but not "avoided deforestation" (Fry 2008). Avoided deforestation returned to the agenda for the post-2012 commitment period, as several developing countries, including Costa Rica and Papua New Guinea, have pushed for it to be included since 2005. Negotiations have been under way since then, and the Conference of the Parties in 2013 has confirmed finalization of the "Warsaw Framework for REDD+" (Corbera et al. 2010).

Part of the enthusiasm for REDD+, particularly from some of the same actors involved for many years in forest policy debates, stems from the potential amounts of funding that could be linked to carbon. Conservationists have argued that REDD+ has the potential for secondary add-on effects (known as co-benefits), such as conservation of biological diversity, leading many to suggest REDD+ is a win–win option (Hagerman et al. 2012). A number of regional and national projects, primarily implemented by bilateral development donors and NGOs, have begun pilot projects to prepare countries for REDD+ implementation in the future (Cerbu et al. 2011). These "REDD+ readiness" programs include the World Bank Forest Carbon Partnership Facility (FCPF) and the United Nations REDD readiness project (UN-REDD).

The possible economic benefits of REDD+ to Vietnam have been estimated at US$60–100 million per year (Ebeling and Yasué 2008), as more than half of Vietnam's total greenhouse gas emissions in the past decade has been attributable to the land-use sector, including agriculture and forestry (Hoa et al. 2012). A new national REDD+ steering committee was established by the government in early 2011, coordinated by the Ministry of Agriculture and Rural Development, and a National REDD+ Network was set up in 2009 for NGOs and donors to offer advice. At least seventeen pilots are under way in assorted provinces to publicize REDD+, conduct carbon baseline measurement, and perform other activities. Like PES projects before them, the REDD readiness pilots are being implemented by NGOs and donors with a wide range of interests, from community-based rural development to biodiversity conservation. The largest REDD+ pilot project has focused on Lam Dong province (site of the above-mentioned PES pilot), where the UN-REDD project is operating, and a phase two will soon expand REDD+ activities to six more provinces (Pham et al. 2012).

Questioning Win–Win Scenarios

Despite often being presented as a win–win scenario that will tackle both conservation and poverty issues, much as other forest projects were supposed to in the past, questions need to be asked as REDD+ activities expand. Can REDD+ projects overcome some of the past barriers to successful forest conservation in Vietnam? Previous challenges to ICDPs, decentralization, and PES approaches have included poor understanding of the most significant drivers of land-use change, with a dominant focus on ethnic minorities and swidden agriculture to the exclusion of attention to corruption, excessive state logging, and industrial cash crop expansion; the regional unevenness of secure land tenure rights to forests, with only 25 percent of forest area under nonstate management; and a lack of support for local households to truly participate in the management of the benefits of forest preservation, such as in community forestry (Sikor and To 2011; Sunderlin and Huynh 2005; To 2009). Will REDD+ be able to do things differently?

To answer this question, anthropological analysis can explore how knowledge is being produced in REDD+ and how climate-related land-use plans compare with previous forest management approaches (Mathews, this volume). For example, which agents of deforestation are REDD+ projects aimed at, and how is this knowledge generated? How are forest communities that are targets of REDD+ designated and represented? Are these agents the true drivers of forest change or simply convenient or more visible scapegoats? Additionally, analysis is needed of the emerging regulatory apparatuses that are associated with REDD+. To what degree are REDD+ projects similar to or different from previous attempts to regulate forest use, particularly in regard to land tenure and participation questions? And are the emergent regulations being proposed truly about climate or about long-standing questions of land-use rights and responsibilities?

Anthropological tools of participant observation in a number of REDD+ readiness meetings and workshops in Vietnam since 2008, as well as field research involving household surveys and interviews in the pilot REDD+ province of Lam Dong since 2011, form the basis of my analysis. Such an approach informs the conclusion that an overattentive focus on technological problems and the need to calculate comparable commodities, such as carbon per unit of forest, and the number of households that have consented to participate, have dominated pilot "REDD+ readiness" activities to date. By presenting REDD+ as a series of checklists and benchmarks that need to be met in order for forestry funding to flow, donors and other actors within Vietnam have neglected a deeper problematization of forest rights and responsibilities, much as previous approaches to forest management did.

Defining Targets of REDD+

Emerging REDD+ projects share with earlier policies a lack of clear understanding of the key agents of deforestation and even common definitions of what deforestation and forests are. For example, in Vietnam's submissions to the FCPF and UN-REDD programs asking for funding, the documents noted that drivers of deforestation "are generally agreed to be a result of: (i) conversion to agriculture (particularly to industrial perennial crops); (ii) unsustainable logging (notably illegal logging); (iii) the impacts of infrastructure development; and (iv) forest fires" (FCPF 2011; UNREDD 2010: 26).[2] Yet on the ground, the main joint government–donor pilot project that has been funded to date, namely, activities by the UN-REDD program in Lam Dong, has concentrated only on smallholder households of ethnic minorities, who have been targeted for awareness-raising activities around REDD+ (interviews with Lam Dong provincial officials, 2012). Other drivers, including infrastructure development (such as roads, hydropower, and industrial parks) and logging, have not received much attention.

For example, at a workshop in January 2014 on the progress of a REDD+ pilot in Dien Bien province, a mountainous area of Vietnam's far northwest that is dominated by ethnic minorities, provincial officials blamed the majority of forest loss in the area on shifting cultivation. They even went so far as to present figures that twenty-two thousand ha of forest had been recently lost to swidden, which contradicted an earlier presentation of satellite data on forest loss, which had estimated only fourteen thousand ha of land conversion. The fact that remotely sensed data for all types of forest conversion had been increased by over 50 percent and then attributed to ethnic minorities indicates deep-rooted inclinations to place blame on some populations but not others. An agricultural extension officer in Lam Dong confirmed these tendencies, noting in an interview that "the main reasons for land conversion are people who have rights, people who have money. But we blame swidden instead" (interview, January 2014).

Further, despite the fact that many of the REDD+ pilots are being targeted at ethnic minority communities, little attention has been paid to the fact that ethnicity is not allowed to be used as the basis for land claims or self-governance in Vietnam (McElwee 2004), and collective ownership of land is still politically difficult, which is one reason why so little land (less than 1 percent of the total forest estate) has been allocated to communities. The state of Vietnam is unlikely ever to allow ethnic minorities as collective groups to have extensive communal land tenure titles. This will likely keep REDD+ in Vietnam from achieving some of the goals of indigenous development that have been advocated at global levels (Dressler et al. 2012; Brown 2013; Lyster et al. 2013).[3]

I myself witnessed how the sensitivities of land rights and ethnicity were skated over in the local pilots through a REDD+ training session in August 2012 in Lam Dong. During an afternoon devoted to understanding the principles of Free, Prior and Informed Consent (FPIC), the young female lecturer from the nearby regional university tried to give examples to the audience of local officials of why FPIC needed to be used and what benefits it could bestow, noting that FPIC was likely to be a requirement of international REDD+ donor funding. Rather than discuss the actual land tenure situation in the province in which we were working, in which no ethnic minorities hold collective titles to land and most ethnic minorities do not hold individual titles to forest land either (titles are available only for agricultural land), she chose an example from Australia. In the case study, Rio Tinto had, in the late 1990s, signed the Yandi Land Use Agreement with the Gumala Aboriginal Corporation, which had been created to represent the collective interests of the traditional owners of the land, and FPIC had been used to get these owners' consent to the development of an ironworks on their land (see Mahanty and McDermott 2013 for a history of FPIC in mining). Ironically, in this example, the Nyiyaparli, Banyjima, and Innawonga peoples who were negotiating with Rio Tinto had ancestral ownership of land, which they were able to use as the basis for incorporation as a firm that could negotiate with the powerful conglomerate. This would be politically unheard of in Vietnam. The confused questions that accompanied the presentation on the Australian FPIC example (such as one local agricultural official who whispered to me, "What is an indigenous corporation?") gave an indication of the difficulty of translating the idea of FPIC from a country with recognition of indigenous peoples and collective land rights to one without. These problems were confirmed in an interview with the head of the provincial agricultural extension service, who had been extensively involved with REDD+ pilots. He noted, "FPIC is taken from a foreign model, and it's not really suitable for Vietnam. FPIC is based on the idea that communities have rights and voice [over forests]. . . . The UN requires certain things like FPIC, and we did it according to the requirements, but what should be the long term way to institutionalize this [participation] in a way that makes sense for Vietnam?" (interview, January 2014).

Expansion of REDD+ Regulation

REDD+ projects in Vietnam have not yet transferred any money to participating localities or imposed new restrictions on forest use. What they have done so far is spend most of their time on getting "REDD+ ready." New checklists of things that must be done to be "ready" are proliferating, including having completed FPIC processes, having undertaken a Participatory Governance Assessment, and having a traceable monitoring, reporting, and verification system for carbon emissions. These REDD+ requirements have largely been devoted to "making things the same" (MacKenzie 2009), that is, making complex ecological systems amenable to a numerical valuation, such as carbon per hectare, which could be used across the country, or using a simple safeguard "tool" to ensure that each and every community that was involved in REDD+ had given consent to their participation through FPIC. As the Lam Dong extension official alluded to previously, these requirements are seen as being outside objectives pushed by the UN that have been difficult for local officials to understand. Trainings are being held in local areas to pass the new global acronyms and requirements downward, and the rapid expansion of these norms has required new specialists familiar with the language. In this manner, REDD+ policies are bringing in forms of what are referred to in the anthropological literature as new "audit cultures" (Kipnis 2008; Strathern 2000). Below I show how these new types of audits, regulations, and checklists were interpreted by Vietnamese actors and how they ran into problems on the ground.

At the REDD+ training I attended in Lam Dong in 2012, the trainer who had used the Australian land rights example mentioned previously had had problems from the start in getting the definition of FPIC across. Speaking in Vietnamese, she had translated *free* as *self-volunteerism* (*su tu nguyen*), *prior* as *before* (*truoc*), *informed* as *enough information* (*du thong tin*) and *consent* as *collective agreement* (*su dong thuan*).[4] (The unclear relationship between self-volunteering and then being part of an agreed-upon consenting collective was not discussed.) The trainer's approach was to try to get across that FPIC had to be implemented prior to any REDD+ work expanding in the province, and that this process would help guarantee both local participation and prevention of abuse. If a project had "done FPIC" prior to implementation of any REDD activities, then it would be considered a success in safeguarding participation. This idea that FPIC was a simple, easily applied tool was widespread. For example, in an interview with a local university official who had been contracted to help with FPIC, he expressed great excitement at the business opportunities it presented; as it was so new, he was going to set up a consulting company to be hired to go "do FPIC." "Doing FPIC" would involve spending a half a day in a community, with two hours for training on what FPIC was, followed by collective voting on whether to allow REDD+ to proceed. The official enthused that if every village in Vietnam had to do an FPIC checklist he would be in business for years with this new firm.

But the effectiveness of FPIC as the sole means to safeguard rights is questionable. In a few pilot UN-REDD tests at the village level, for example, meetings to get consent for the development of REDD+ were held but were very short, no more than a couple of hours in length, and only forty-five minutes were allocated for questions and answers after the "awareness-raising activities" (mainly explanations of what climate change is and how forests mitigate carbon). Then villagers had to make the decision to consent or not consent to REDD+ activities, often with open shows of hands. No one in the FPIC preliminaries was presented with any information on the possible risks and costs of participation (that is, changes in agricultural practices that might need to be made). Rather, the participants were asked general questions like, "Do you want your forests to be conserved?" (Nguyen et al. 2010). Not surprisingly, most people supported REDD+, since the question did not refer to any costs that might be incurred in forest conservation or how it might be carried out.

Further, the FPIC process in Lam Dong has been mostly undertaken in villages that do not have formal land tenure rights to nearby forest lands, which are actually owned by the state. Thus the villages who were "consulted" and who "consented" to join REDD+ have no formal landownership rights granted to them that would enable them to have legally defensible say over local forest management. UN-REDD project documents on FPIC in Vietnam have been widely promoted as the first example of a UN-REDD project anywhere to carry out FPIC, but they do not discuss the problem that the majority of households were voting to participate in REDD+ on lands that were not really theirs to offer consent for (UN-REDD 2010a). While the pilot community met all the requirements on the checklist for FPIC (that is, they were given information, it happened prior to REDD+ implementation, and they voted voluntarily), can such a system truly be considered to be legal consent over forest activities and regulation? Interviews with people who had participated in the FPIC sessions a few months after the fact revealed that many had no idea what they had raised their hands for, and most had forgotten the idea of the meeting altogether (personal communication, UN-REDD consultant, 2013). In fact, in one commune that has seen numerous consultants going in and out, doing FPIC meetings and carbon assessments and other readiness activities, but providing no actual financial support for local households, a local leader had coopted the REDD acronym, stating that in Vietnamese it actually stood for "Roi Em Den va Di," which translates to "Here you come and go again" (interview, Forest Department of Lam Dong, 2014).

A second type of checklist-ification that was occurring in Vietnam involved the work of making forest carbon in diverse areas comparable to one another through the mapping of carbon distributions, the identification of tree types that stored more carbon than others, and the development of a formula to determine payment rates for protection of carbon-rich forests. Consultants have been hired to create lists of tree allometric equations, which are an expression of the amount of carbon likely to be found in particular species of tree, based on average stand density, wood volume, wood density, bark-to-wood ratio, and other factors (UN-REDD 2013). Local valuations and assessments of forests' usefulness were not needed for such equations, which seem to fit the definition proposed by Michel Callon of "dispositifs de calcul," or calculative mechanisms, that form distinctions between things and shape the parameters of new understandings of calculated entities (Callon and Muniesa 2005; Callon 2009). In this imagining, the calculative mechanisms turned forests from ecological groupings of trees into stocks of carbon.

These extra-local calculative mechanisms were not accepted without challenges, however, as the use of carbon content to value trees also raises certain ontological questions: what is a tree? If a tree is simply a mass of woody material made of carbon, as the tree allometric equations suggested, then many species that have long been classified in Vietnam as agricultural crops, such as coffee, cashew, and rubber, potentially become new actors in REDD+ plans. Indeed, households interviewed in our local research sites raised these questions themselves: once told of what carbon was (using the term *cax bon* in Vietnamese, adopted from English, as there was no previous word for it), several households said that they should be eligible to receive carbon payments for coffee trees, the very planting of which had driven figures on deforestation that made Lam Dong province a site for REDD+ targeting. Why shouldn't coffee be considered carbon under local reasoning, if *cax bon* was simply visible woody matter? Similarly, several local provinces in areas where rubber grows have argued that replacing poor quality natural forest with rubber trees is not a significant change in ecotype, as it is just substituting one type of carbon-holding woody material with another. These arguments have recently been boosted by a decree of the prime minister which allows rubber conversion with central government permission. Given that industrial rubber expansion in Southeast Asia has been implicated in major patterns of deforestation in recent years (Fox and Castella 2013), it is ironic that the ontological questions raised by REDD+ of what counts as carbon-worthy trees may in fact further drive local deforestation.

COMPARING NEW FOREST POLICY WITH OLD

The saying "old wine in new bottles" is widespread in Vietnam (*binh moi, ruou cu*) and is often applied to the numerous attempts that have occurred over the past twenty years to improve the forest sector, from land allocation to reforestation funding to market-based conservation. REDD+ may indeed be the newest bottle, but it must be seen in the light of a long history of global interest in forest management in the tropics. Despite years of searching for outcomes that do not exacerbate inequity and poverty, no magic bullet has been found to deal with the complicated issues of conservation and development that forest management entails, and this dilemma can be seen in the history of forest management in Vietnam. In this respect, REDD+ faces the same challenges that troubled earlier panacea approaches like ICDPs and community-based natural resource management. Identifying REDD+ solely as a new type of global climate governance hides this long history of struggles over local forest rights, which this chapter has sought to contextualize.

Whether or not REDD+ will be able to make a difference in global carbon emissions remains an open question. As evidenced by the case study in Vietnam, much of the production of knowledge about forests and people under REDD+ and the regulations that are developing seem to be mostly a rehashing of previously unsuccessful endeavors. Like other approaches before it, REDD+ may not be focused on the right forest change drivers, as local forest users, many of whom have long been excluded from formal land title and recognition of their forest use, continue to be the targets of environmental rule and intervention, while other agents of deforestation like export agriculture, migration, and infrastructure development are ignored. Similar processes of overattention to local actors at the expense of global drivers can be noted in other REDD+ pilot sites (Leggett and Lovell 2012; Murdiyarso et al. 2012; Pokorny et al. 2013; Awono et al. 2014). Given this uneven focus, exploring how various actors make assessments of the drivers of deforestation or climate emissions can be a highly useful contribution of anthropologists (e.g., see Milne 2012; Leggett and Lovell 2012; Bull 2013).

Anthropologists also have a crucial role to play in drawing attention to the social justice and equity implications of environmental interventions. Such questions require us to look carefully at the globalization of forest and climate policies like REDD+, given the specificity of social and property relations in most countries (Forsyth and Sikor, 2013; Evans et al. 2014). In Vietnam, REDD+ projects have not been designed to deal with two major equity issues that have long plagued the forest sector: uneven land tenure and the lack of a strong participatory role for local people, especially ethnic minorities, in forest management. REDD+ pilots have appeared to steer away from sensitive issues such as ethnicity and land tenure rights, instead stressing more bureaucratic and apolitical calculative checklist approaches for supposed participation through FPIC and environmental protection through carbon equations. While these processes of checklist-ification have distilled complexity into neat formulaic simplicity, they have prevented a deeper discussion of fundamental, often political, issues related to land tenure and other rights. As local actors noted, the use of FPIC as a tool has been primarily about meeting the demands of global donors' requirements, not about a fundamental reassessment of land rights and participation in forest management.

It is not clear if REDD+ has been cast in climate solution terms precisely to avoid answering these difficult questions about rights and citizen participation to begin with. Presenting REDD+ as an urgent project for climate emissions reductions and as "low-hanging fruit" in global climate negotiations (Laurance 2008) may serve to gloss over the eminently political implications of a restructuring of forest rights and responsibilities. Certainly in Vietnam, REDD+ runs the risk of simply replicating existing patterns of institutionalized management of land and commodities that are spatially uneven and socially unequal. Similar conclusions have been drawn for other climate mitigation approaches, indicating a strong role for anthropologists to continue to raise important concerns regarding participation and equity.

NOTES

The research for this chapter was made possible by a grant from the National Science Foundation Geography and Regional Science Division for the project "Downscaling REDD policies in developing countries: Assessing the impact of carbon payments on household decision-making and vulnerability to climate change in Vietnam" (grant #11028793). My Vietnamese collaborators on this project were additionally supported by a U.S. Agency for International Development (USAID) Partnerships for

Enhanced Engagement in Research grant: "Research and capacity building on REDD+, livelihoods, and vulnerability in Vietnam: developing tools for social analysis of development planning." I would like to thank my Vietnamese collaborators who contributed to the collection of data outlined in this chapter: Nghiem Phuong Tuyen, Le Thi Van Hue, Tran Huu Nghi, Vu Thi Dieu Huong, Ha Thi Thu Hue, Nguyen Viet Dung, and Nguyen Xuan Lam.

1. REDD is now often referred to as REDD+ to indicate the additionality of "the role of conservation, sustainable management of forests, and enhancement of forest carbon stocks in developing countries." Others have also suggested REDD++, indicating the "need to examine emissions from all land-use activities in efforts to reduce terrestrial emissions" (Agrawal et al. 2011). REDD+ is used here to indicate the current state of the suggested program (Pistorius 2012).
2. A review of preparatory documents for the global Forest Carbon Partnership Facility indicates that many national authorities, not just Vietnam, have pinned blame for deforestation on simplistic culprits like swidden agriculture rather than on the complicated processes that are often at work, including tenure issues, competing land uses, and corruption (Dooley et al. 2008).
3. Indeed, the lack of attention to political issues of indigenous rights and land tenure by REDD+ in Vietnam and other states was recently blasted by an advocacy organization at the UN Permanent Forum for Indigenous Peoples (Goldtooth 2013).
4. The acronym for FPIC in Vietnamese would have ended up being STNTTDTTSDT, so everyone simply used FPIC, despite the fact that there is no letter *F* in the Vietnamese language.

REFERENCES CITED

Agrawal, A., D. Nepstad, and A. Chhatre. 2001. "Reducing Emissions from Deforestation and Forest Degradation." *Annual Review of Environment and Resources* 36, no. 1: 373–96.

Awono, A., O. A. Somorin, R. E. Atyi, and P. Levang. 2014. "Tenure and Participation in Local REDD+ Projects: Insights from Southern Cameroon." *Environmental Science and Policy* 35, no. 1: 76–86.

Baviskar, A. 2000. "Claims to Knowledge, Claims to Control: Environmental Conflict in the Great Himalayan National Park, India." In *Indigenous Environmental Knowledge and Its Transformations: Critical Anthropological Perspectives*, ed. A. Bicker, R. Ellen, and P. Parkes, 101–19. Reading, UK: Harwood Academic Publishers.

Brandon, K. E., and M. Wells. 1992. "Planning for People and Parks: Design Dilemmas." *World Development* 20, no. 4: 557–70.

Brosius, J. P. 2002. "Endangered Forest, Endangered People: Environmentalist Representations of Indigenous Knowledge." *Human Ecology* 25: 47–69.

———. 2006. "Common Ground between Anthropology and Conservation Biology." *Conservation Biology* 20, no. 3: 683–85.

Brown, M. 2013. *Redeeming REDD: Policies, Incentives and Social Feasibility for Avoided Deforestation.* Abingdon, UK: Earthscan.

Bull, C. 2013. "Overlapping Spaces: The Politics of REDD in Action – An Anthropological Account from the Bolivian Amazon." MS thesis, University of Oslo.
Callon, M. 2009. "Civilizing Markets: Carbon Trading Between in Vitro and in Vivo Experiments." *Accounting* 34: 535–48.
Callon, M., and F. Muniesa. 2005. "Peripheral Vision: Economic Markets as Calculative Collective Devices." *Organization Studies* 26, no. 8: 1229–50.
Cerbu, G. A., B. M. Swallow, and D. Y. Thompson. 2011. "Locating REDD: A Global Survey and Analysis of REDD Readiness and Demonstration Activities." *Environmental Science and Policy* 14, no. 2:168–80.
Clement, F., and J. M. Amezaga. 2009. "Afforestation and Forestry Land Allocation in Northern Vietnam: Analysing the Gap Between Policy Intentions and Outcomes." *Land Use Policy* 26, no. 2: 458–70.
Coe, C. A. 2012. "Local Power Structures and Their Effect on Forest Land Allocation in the Buffer Zone of Tam Dao National Park, Vietnam." *Journal of Environment and Development* 20, no. 10: 1–30.
Colfer, C., and D. Capistrano, eds. 2012. *The Politics of Decentralization: Forests, Power and People.* London: Routledge.
Corbera, E., and H. Schroeder. 2011. "Governing and Implementing REDD+." *Environmental Science and Policy* 14, no. 2: 89–99.
Corbera, E., and K. Brown. 2010. "Offsetting Benefits? Analyzing Access to Forest Carbon." *Environment and Planning A* 42, no. 7: 1739–61.
Corbera, E., M. Estrada, and K. Brown. 2010. "Reducing Greenhouse Gas Emissions from Deforestation and Forest Degradation in Developing Countries: Revisiting the Assumptions." *Climatic Change* 100, no. 3: 355–88.
De Koninck, R. 1999. *Deforestation in Viet Nam.* Ottawa: IDRC.
Doane, M. 2007. "The Political Economy of the Ecological Native." *American Anthropologist* 109, no. 3: 452–62.
Déry, S., and R. Vanhooren. 2011. "Protected Areas in Mainland Southeast Asia, 1973–2005: Continuing Trends." *Singapore Journal of Tropical Geography* 32, no. 2: 185–202.
Dooley, K., T. Griffiths, H. Leake, and S. Ozinga. 2008. "Cutting Corners: World Bank's Forest and Carbon Fund Fails Forests and Peoples." Moreton in Marsh, UK: FERN and Forest Peoples Programme.
Dove, M. R. 1983. "Theories of Swidden Agriculture and the Political Economy of Ignorance." *Agroforestry Systems* 1, no. 2: 85–99.
———. 1995. "The Theory of Social Forestry Intervention: The State of the Art in Asia." *Agroforestry Systems* 30, no. 3: 315–40.
———. 2003. "Forest Discourses in South and Southeast Asia: A Comparison with Global Discourses." In *Nature in the Global South: Environmental Projects in South and Southeast Asia*, ed. P. Greenhough and A. Tsing, 103–23. Durham: Duke University Press.
Dressler, W. H., M. H. McDermott, W. Smith, and J. Pulhin. 2012. "REDD Policy Impacts on Indigenous Property Rights Regimes on Palawan Island, the Philippines." *Human Ecology* 40, no. 5: 679–91.
Ebeling, J., and M. Yasué. 2008. "Generating Carbon Finance Through Avoided Deforestation and Its Potential to Create Climatic, Conservation and Human

Development Benefits." *Philosophical Transactions of the Royal Society B: Biological Sciences* 363, no. 1498: 1917–24.

Evans, K., L. Murphy, and W. de Jong. 2014. "Global Versus Local Narratives of REDD: A Case Study from Peru's Amazon." *Environmental Science and Policy* 35: 98–108.

FCPF. 2011. "Synthesis Review of R-PP of Vietnam PDR." Washington: Forest Carbon Partnership Facility.

Forest Inventory and Planning Institute. 1996. *Final Report on Forest Resource Changes (1991–1995).* Hanoi: Forest Inventory and Planning Institute.

Forsyth, T. J., and A. Walker. 2008. *Forest Guardians, Forest Destroyers: The Politics of Environmental Knowledge in Northern Thailand.* Seattle: University of Washington Press.

Forsyth, T. J., and T. Sikor. 2013. "Forests, Development and the Globalisation of Justice." *Geographical Journal* 179: 114–21.

Fox, J., and J. C. Castella. 2013. "Expansion of Rubber (*Hevea brasiliensis*) in Mainland Southeast Asia: What Are the Prospects for Smallholders?" *Journal of Peasant Studies* 40, no. 1: 155–70.

Fry, I. 2008. "Reducing Emissions from Deforestation and Forest Degradation: Opportunities and Pitfalls in Developing a New Legal Regime." *Review of European Community and International Environmental Law* 17, no. 2: 166–82.

Gilmour, D. A., and V. S. Nguyen. 1999. "Buffer Zone Management in Vietnam." Hanoi: International Union for the Conservation of Nature.

Goldtooth, T. 2013. United Nations Permanent Forum for Indigenous Peoples 12th Session May 18–31. Joint Collective Statement by Bruce "Tom" Goldtooth, Indigenous Educational Network-Indigenous Environmental Network with Supportive Sign-On of the Seventh Generation Fund, American Indian Law Alliance and Tonatierra. New York.

GSO. 2007. *Results of the 2006 Rural, Agricultural and Fishery Census.* Volume 3. Hanoi: General Statistical Office, Vietnam.

Gullison, R. E., P. C. Frumhoff, J. G. Canadell, C. B. Field, D. C. Nepstad, K. Hayhoe, R. Avissar, L. M. Curran, P. Friedlingstein, C. D. Jones, and C. Nobre. 2007. "Tropical Forests and Climate Policy." *Science* 316, no. 5827: 985.

Haenn, N. 1999. "The Power of Environmental Knowledge: Ethnoecology and Environmental Conflicts in Mexican Conservation." *Human Ecology* 27, no. 3: 477–91.

———. 2002. "Nature Regimes in Southern Mexico: A History of Power and Environment." *Ethnology* 41, no. 1: 1–26.

Hagerman, S., R. Witter, C. Corson, D. Suarez, E. M. Maclin, M. Bourque, and L. Campbell. 2012. "On the Coattails of Climate? Opportunities and Threats of a Warming Earth for Biodiversity Conservation." *Global Environmental Change* 22, no. 3: 724–35.

Hoa, N. T., T. Hasegawa, and Y. Matsuoka. 2012. "Climate Change Mitigation Strategies in Agriculture, Forestry and Other Land Use Sectors in Vietnam." *Mitigation and Adaptation Strategies for Global Change* 19, no. 1: 15–32.

Hoang, M. H., T. H. Do, M. T. Pham, M. van Noordwijk, and P. A. Minang. 2012. "Benefit Distribution across Scales to Reduce Emissions from Deforestation and Forest Degradation (REDD+) in Vietnam." *Land Use Policy 31:* 48–60.

Kipnis, A. B. 2008. "Audit Cultures: Neoliberal Governmentality, Socialist Legacy, or Technologies of Governing?" *American Ethnologist* 35, no. 2: 275–89.

Laungaramsri, P. 2002. *Redefining Nature: Karen Ecological Knowledge and the Challenge to the Modern Conservation Paradigm.* Chennai: Earthworm Books.

Laurance, W. F. 2008. "Can Carbon Trading Save Vanishing Forests?" *BioScience* 58, no. 4: 286–87.

Leggett, M., and H. Lovell. 2012. "Community Perceptions of REDD+: A Case Study from Papua New Guinea." *Climate Policy* 12, no. 1: 115–34.

Lyster, R., C. MacKenzie, and C. L. McDermott. 2013. *Law, Tropical Forests and Carbon: The Case of REDD.* Cambridge: Cambridge University Press.

MacKenzie, D. 2009. "Making Things the Same: Gases, Emission Rights and the Politics of Carbon Markets." *Accounting, Organizations and Society* 34, no. 3–4: 440–55.

Mahanty, S., and C. McDermott. 2013. "How Does 'Free, Prior and Informed Consent' (FPIC) Impact Social Equity? Lessons from Mining and Forestry and Their Implications for REDD." *Land Use Policy* 35: 406–16.

Mathews, A. S. 2008. "State Making, Knowledge, and Ignorance: Translation and Concealment in Mexican Forestry Institutions." *American Anthropologist* 110, no. 4: 484–94.

McElwee, P. D. 2002. "Lost Worlds and Local People." In *Conservation and Mobile Indigenous Peoples: Displacement, Forced Settlement, and Sustainable Development,* ed. D. Chatty and Marcus Colchester, 312–29. Oxford: Berghahn Books.

———. 2004. "Becoming Socialist or Becoming Kinh? Government Policies for Ethnic Minorities in the Socialist Republic of Viet Nam." In *Civilizing the Margins: Southeast Asian Government Policies for the Development of Minorities,* 182–213. Ithaca: Cornell University Press.

———. 2005. "You Say Illegal, I Say Legal: The Relationship Between 'Illegal' Logging and Land Tenure, Poverty, and Forest Use Rights in Vietnam." *Journal of Sustainable Forestry* 19, nos. 1/2/3: 97–135.

———. 2009. "Reforesting 'Bare Hills' in Vietnam: Social and Environmental Consequences of the 5 Million Hectare Reforestation Program." *AMBIO: A Journal of the Human Environment* 38, no. 6: 325–33.

———. 2010. "Resource Use Among Rural Agricultural Households Near Protected Areas in Vietnam: The Social Costs of Conservation and Implications for Enforcement." *Environmental Management* 45, no. 1: 113–31.

———. 2012. "Payments for Environmental Services as Neoliberal Market-Based Forest Conservation in Vietnam: Panacea or Problem?" *Geoforum* 43, no. 3: 412–26.

———. forthcoming. *Forests Are Gold: Trees, People and Environmental Rule in Vietnam.* Seattle: University of Washington Press.

McElwee, P. D., T. Nghiem, H. Le, H. Vu, and N. Tran. 2014. "Payments for Environmental Services and Contested Neoliberalisation in Developing Countries: A Case Study from Vietnam." *Journal of Rural Studies,* 2014 online first.

Meyfroidt, P., and E. F. Lambin. 2008. "Forest Transition in Vietnam and Its Environmental Impacts." *Global Change Biology* 14, no. 6: 1319–36.

———. 2009. "Forest Transition in Vietnam and Displacement of Deforestation Abroad." *PNAS* 106, no. 38: 16139–144.

Milne, S. 2012. "Grounding Forest Carbon: Property Relations and Avoided Deforestation in Cambodia." *Human Ecology* 40, no. 5: 693–706.

Murdiyarso, D., M. Brockhaus, W. D. Sunderlin, and L. Verchot. 2012. "Some Lessons Learned from the First Generation of REDD+ Activities." *Current Opinion in Environmental Sustainability* 4, no. 6: 678–85.

Ngo, D. T., et al. 2006. "Cam Nang Nganh Lam Nghiep: Chuong Quan Ly Lam Truong Quoc Doanh." Hanoi: Ministry of Agriculture and Rural Development.

Nguyen, Q. T. 2011. "Payment for Environmental Services in Vietnam: An Analysis of the Pilot Project in Lam Dong Province." *Institute for Global Environmental Strategies Forest Conservation, Livelihoods, and Rights Project.* Kanagawa, Japan: Institute for Global Environmental Strategies.

Nguyen, Q. T., et al. 2010. "Evaluation and Verification of the Free, Prior and Informed Consent Process Under the UN-REDD Programme in Lam Dong Province, Vietnam." Bangkok: RECOFTC.

Peluso, N. L., and P. Vandergeest. 2001. "Genealogies of the Political Forest and Customary Rights in Indonesia, Malaysia, and Thailand." *Journal of Asian Studies* 60, no. 3: 761–812.

Pham, T. T. T., B. M. Campbell, and S. Garnett. 2009. "Lessons for Pro-Poor Payments for Environmental Services: An Analysis of Projects in Vietnam." *Asia Pacific Journal of Public Administration* 31, no. 2: 117–33.

Pham, T. T. T., M. Moeliono, Nguyen Thi Hien, Nguyen Huu Tho, and Vu Thi Hien. 2012. "The Context of REDD+ in Vietnam: Drivers, Agents and Institutions." Bogor: Center for International Forestry Research (CIFOR).

Pham, T. T. T., K. Bennett, Vu Tan Phuong, J. Brunner, Le Ngoc Dung, and Nguyen Dinh Tien. 2013. "Payments for Forest Environmental Services in Vietnam: From Policy to Practice." Bogor: Center for International Forestry Research (CIFOR).

Phelps, J., E. L. Webb, and A. Agrawal. 2010. "Does REDD+ Threaten to Recentralize Forest Governance?" *Science* 328, no. 5876: 312.

Pistorius, T. 2012. "From RED to REDD+: The Evolution of a Forest-Based Mitigation Approach for Developing Countries." *Current Opinion in Environmental Sustainability* 4, no. 6: 638–45.

Pokorny, B., I. Scholz, and W. de Jong. 2013. "REDD+ for the Poor or the Poor for REDD+? About the Limitations of Environmental Policies in the Amazon and the Potential of Achieving Environmental Goals through Pro-Poor Policies." *Ecology and Society* 18, no. 2: 3.

Polet, G. 2003. "Co-Management in Protected Areas: The Case of Cat Tien National Park, Southern Vietnam." In *Co-Management of Natural Resources in Asia: A Comparative Perspective*, ed. Gerard A. Persoon, Diny M. E. Van Est, and Percy E. Sajise, 25–42. Copenhagen: NIAS Press.

Raintree, J. 2004. "Domesticated NTFPs, Secured Livelihoods: Impacts of NTFP Domestication and Agro-Forestry on Poverty Alleviation and Livelihood Improvement." Hanoi: IUCN.

Rugendyke, B., and N. T. Son. 2005. "Conservation Costs: Nature-Based Tourism as Development at Cuc Phuong National Park, Vietnam." *Asia Pacific Viewpoint* 46, no. 2: 185–200.

Service scientifique de l'Agence économique de l'Indochine. 1931. "La Situation Forestière de l'Indochine." Paris: Exposition Coloniale Internationale.

Sikor, T., and N. T. Tran. 2007. "Exclusive Versus Inclusive Devolution in Forest Management: Insights from Forest Land Allocation in Vietnam's Central Highlands." *Land Use Policy* 24, no. 4: 644–53.

Sikor, T., and Q. T. Nguyen. 2007. "Why May Forest Devolution Not Benefit the Rural Poor? Forest Entitlements in Vietnam's Central Highlands." *World Development* 35, no. 11: 2010–25.

Sikor, T., and X. P. To. 2011. "Illegal Logging in Vietnam: *Lam Tac* (Forest Hijackers) in Practice and Talk." *Society and Natural Resources* 24, no. 7: 688–701.

Sivaramakrishnan, K. 2000. "State Sciences and Development Histories: Encoding Local Forestry Knowledge in Bengal." *Development and Change* 31: 61–89.

Sowerwine, J. C. 2004. "Territorialisation and the Politics of Highland Landscapes in Vietnam: Negotiating Property Relations in Policy, Meaning and Practice." *Conservation and Society* 2, no. 1: 97–136.

Spelchan, D., et al. 2011. "Co-Management/Shared Governance of Natural Resources and Protected Areas in Viet Nam." Soc Trang: Deutsche Gesellschaft für Internationale Zusammenarbeit (GIZ) GmbH.

Strathern, M., ed. 2000. *Audit Cultures: Anthropological Studies in Accountability, Ethics, and the Academy.* London: Routledge.

Sunderlin, W. D., and T. B. Huynh. 2005. *Poverty Alleviation and Forests in Vietnam.* Bogor: CIFOR.

Thomas, F. 2009. "Protection des Forêts et Environnementalisme Coloniale: Indochine 1860–1945." *Revue d'Histoire Moderne et Contemporaine* 56e, no. 4: 104–36.

Thuan, D. D., et al. 2006. *Forestry, Poverty Reduction and Rural Livelihoods in Vietnam.* Hanoi: Labour and Social Affairs Publishing House.

To, X. P. 2009. "Why Did the Forest Conservation Policy Fail in the Vietnamese Uplands? Forest Conflicts in Ba Vi National Park in Northern Region." *International Journal of Environmental Studies* 66, no. 1: 59–68.

To, X. P., W. H. Dressler, S. Mahanty, T. T. T Pham, and C. Zingerli. 2012. "The Prospects for Payment for Ecosystem Services (PES) in Vietnam: A Look at Three Payment Schemes." *Human Ecology* 40, no. 2: 237–49.

Tran, N. T., and T. Sikor. 2006. "From Legal Acts to Actual Powers: Devolution and Property Rights in the Central Highlands of Vietnam." *Forest Policy and Economics* 8, no. 4: 397–408.

Tuynh, V. H., and P. X. Phuong. 2001. "Impacts and Effectiveness of Logging Bans in Natural Forests: Vietnam." In *Forests Out of Bounds: Impacts and Effectiveness of Logging Bans in Natural Forests in Asia-Pacific,* ed. P. Durst, T. R. Waggoner, T. Enters, and T. L. Cheng, 185–207. Bangkok: Food and Agriculture Organization Regional Office for Asia and the Pacific.

UN-REDD Vietnam. 2009. "About Pilot Site: Di Linh and Lam Ha Districts, Lam Dong Province." Hanoi: UN-REDD.

———. 2010a. "Applying the Principle of Free, Prior and Informed Consent in the UN-REDD Programme in Viet Nam." Hanoi: UN-REDD.

———. 2010b. "UN-REDD Viet Nam Programme Phase II: Operationalising REDD+ in Viet Nam." Hanoi: UN-REDD Vietnam.

———. 2012. "Piloting Local Decision Making in the Development of a REDD+ Compliant Benefit Distribution System for Viet Nam." Hanoi: UN-REDD.

———. 2013. "Tree Allometric Equation Development for Estimation of Forest Above-Ground Biomass in Viet Nam: Part A—Introduction and Background of the Study." Hanoi: UN-REDD and Food and Agriculture Organization.

UN-REDD Vietnam and MARD. 2010. "Design of a REDD Compliant Benefit Distribution System for Viet Nam." Hanoi: UN-REDD and Ministry of Agriculture and Rural Development.

Visseren-Hamakers, I. J., C. McDermott, M. J. Vijge, and B. Cashore. 2012. "Trade-Offs, Co-Benefits and Safeguards: Current Debates on the Breadth of REDD+." *Current Opinion in Environmental Sustainability* 4, no. 6: 646–53.

West, P., J. Igoe, and D. Brockington. 2006. "Parks and Peoples: The Social Impact of Protected Areas." *Annual Reviews in Anthropology* 35, no. 1: 251–77.

Wunder, S., S. Engel, and S. Pagiola. 2008. "Taking Stock: A Comparative Analysis of Payments for Environmental Services Programs in Developed and Developing Countries." *Ecological Economics* 65, no. 4: 834–52.

Zingerli, C. 2005. "Colliding Understandings of Biodiversity Conservation in Vietnam: Global Claims, National Interests, and Local Struggles." *Society and Natural Resources* 18, no. 8: 733–47.

Part Two **Knowing Climate Change**

Chapter 4 **Glacial Dramas: Typos, Projections, and Peer Review in the Fourth Assessment of the Intergovernmental Panel on Climate Change**

Jessica O'Reilly

The Intergovernmental Panel on Climate Change (IPCC) is a group of scientist authors who are nominated by their governments and asked to prepare summaries of climate science research. This work underpins the scientific consensus on anthropogenic climate change and was acknowledged in 2007 with a Nobel Peace Prize, awarded jointly to the IPCC and Al Gore. The work of scientific assessments takes place in what Sheila Jasanoff calls the "gray area between science and policy" (1990: 216)—that is, the IPCC produces policy-relevant science advice. But it is difficult to remain impartial in the face of the utopian and dystopian futures afforded by the assessment reports.

If you search for images of the IPCC, you will find hundreds of charts and graphs with a few photographs of people mixed in. The charts and graphs, including the color selection, represent a deep concern to the aesthetics of climate data as convincing and reputable (Mahony 2014a). Those photographs with people are usually shots of large groups taken during author meetings or pictures of meetings, featuring the backs of people's heads. The people face the projector—their work is the production of charts, graphs, and text that assess the state of climate science and projections of future climate change through the year 2100.

As an anthropologist studying how climate science is produced, I find the IPCC to be a challenging organization to think about culturally. To do so, I work with the people who make decisions about the charts, graphs, and text that formally represent the IPCC. While the organization is peopled and placed, with predictable involvement from state governments, NGOs, and universities, the IPCC also performs a sort of cultural erasure. This erasure takes place because of, first, the disciplined scientific writing practice of removing oneself from one's research, in an attempt to limit bias, and second, the Latourian notion that when one tracks scientific publications all the way to their barest essence—to data, numbers, and citations—it is not people who are revealed but nature itself (Latour 1990). Citations also point to networks that the reference is embedded in, shaping objects and subjects in a particular way, glossed over with the veneer of nature (Choy 2011). The main project of the IPCC is to distill the styles and magnitude in which the earth's climate is changing, to read the fate of the planet and its populations from the data that scientists generate, to lay bare the workings of nature so that decision makers can act (see also Mathews, this volume).

My task as an ethnographer of the IPCC is to trace the cultural practices that lead to this distillation of nature as climate change. What follows below is a small dissection of a climate publication and its citations, which I performed by analyzing archived comments on the rough drafts of the Fourth Assessment Report, which was published in 2007. This dissection is anchored in anthropological understandings of texts and documents as culturally produced, explicitly negotiated, often to the minutest detail, and encapsulating particular discursive logics (Ferguson 1990; Riles 1998).

All of the assessment reports of the IPCC have been increasingly scrutinized by climate skeptics for error and fraud (see also Lahsen, this volume). I focus here on two numbers from the Fourth Assessment Report that have become a source of controversy. In late 2009, IPCC watchdogs found an error on the melt rate for the mountain glaciers of the Himalayas. In the report, the authors suggested that the glaciers would disappear by 2035. This number—which suggests an alarmingly fast melt rate and equally alarming changes for local residents—was accompanied by an appropriate-looking citation. However, if one tracks the citation along with earlier drafts of the IPCC chapter, a story unravels about accuracy, expertise, and peer review. By contrast, in my second example, assessors working to estimate global sea level rise in the fourth IPCC report decided to eliminate contributions from the rapid disintegration of ice sheets from Antarctica and Greenland in their projections, thereby making the numbers appear to indicate a decrease over the previous report's estimated sea level rise. The exclusion of this particularly significant contribution to sea level rise caused controversy and confusion among assessors and in the media interpretation of the projection. In the draft review comments, diverse scientists noted this problem and railed against the characterization of sea level rise that eventually was published.

This chapter describes the citational chain along which each reference—there are thousands in an IPCC report—must travel and the ways in which "trust in numbers" (Porter 1995) and "trust in scientists" (Shapin 1995) are supported and undermined. In comparing these two numbers from the IPCC report—numbers of different scales, both in their prominence in the report as well as in potential human impacts—I analyze the challenges that confront this large-scale, international assessment of climate change research. I argue that the IPCC assessment report occasionally contains "casual numbers," that is, numbers not supported by formal, peer-reviewed literature. I also discuss the phenomenon of assessment itself, positing that it moves from a scientific project to one of auditing and accountability. The massive project of contemporary international climate assessment produces knowledge about the state of climate science: the wobbles in the assessment project highlight the contingent nature of this knowledge production.

IPCC ORIENTATION

The projections of the IPCC assessment reports are usually constructed through model runs containing code based upon state-of-the-art climate science. These projections—ranges of numbers that predict events like sea level rise and desertification—are imbued with the labor of thousands of people, millions of dollars of grant money, and peer-reviewed scientific expertise.

Each IPCC assessment report contains three volumes written by corresponding working groups. Working Group I assesses the "Physical Science Basis" of climate change. This working group is overwhelmingly composed of natural and other physical scientists assessing observed measurements, reconstructions of the paleoclimate, and the climate models that project future global climate changes. Working Group II is responsible for assessing

"Impacts, Adaptation, and Vulnerability." Authors in this group—social scientists alongside natural scientists—apply the scientific findings of the first working group onto the various regions of the world and discuss the impacts that climate changes will have on various local ecosystems and human populations. Working Group III specializes in mitigation, in which social scientists, with heavy representation from economists, assess various strategies for slowing, reversing, or rectifying climate changes in various locations and scenarios. Because of differences in content and disciplines as well as in institutional culture, each working group has particular personalities, tones, and traditions. Working Group I in particular separates itself from what some consider the more value-laden second and third working groups.

Preparing an IPCC assessment report takes several years and multiple iterations of drafts. A Technical Support Group assists the scientists through the assessment process. A committee, attending an organizational meeting, sets the table of contents for the assessment report; national governments then nominate potential authors for each chapter. IPCC assessment reports are first reviewed as Zero Order Drafts, which are passed around by the chapter authors to a few key experts. These Zero Order Drafts are not publicly available.

The review process becomes more formal with the First Order Drafts and Second Order Drafts. The First Order Draft undergoes a peer review by experts in the field related to the chapter, expanding upon the self-selected community who reviews the Zero Order Drafts. Anyone can sign up to be a peer reviewer on the IPCC's website (www.ipcc.ch), so reviewers run the gamut from academic scientists to interested nonprofessionals and so-called skeptics.

The Second Order Drafts undergo two sets of reviews: another round of expert reviews and governmental reviews by officials representing various agencies. The writing group authors respond to the comments and finalize the report. A subgroup of authors and IPCC staff draft the Summary for Policymakers, which must be approved by consensus in a plenary meeting of the involved governments. The summary is the only piece of the IPCC assessment reports which is an international, government consensus document; the chapters are consensus documents only among the chapter authors.

By looking at the drafts and the comments associated with them, we can discern the moments of scrutiny given to the two figures I focus on here: the erroneous Himalayan glacier claim and the problematic sea level rise predictions.

OVERSIGHT, REVIEW, AND ACCOUNTABILITY

At first glance, the most appropriate mechanism for a scientific assessment like the IPCC reports would appear to be peer review. Peer review is a complex endeavor built upon an amorphous excellence, belief in the process as generally fair and accurate, and a reliance on the social fragility of consensus building (Lamont 2009). The tradition of peer review as the best, however flawed, means for evaluating new scientific research remains generally upheld.

However, the idea that scientific assessments should undergo quality control via peer review presupposes that a scientific assessment has much in common with a research publication. As assessments have become increasingly formal, international, and institutional, their review has followed suit. The IPCC, arguably the most comprehensive scientific assessment to date, employs peer review, calling it expert review as described above. The practice of peer review in the IPCC, however, is only one of a bundle of oversight mechanisms that also include governmental review and external, institutional review of the IPCC itself.

In addition to reviews by outsiders of the text and practices of the IPCC that mimic the collegiality and subject matter of peer review, IPCC oversight has informally taken on a tone of accountability, even auditing. The reasons for this are twofold: (1) a scientific assessment is not a scientific research publication and therefore requires specific oversight mechanisms including but not limited to peer review, and (2) besides formal oversight arranged by people within the IPCC, the content of the assessment reports is publicly and politically fraught, contested, and scrutinized.

Part of the reviewers' work is to scan texts for inaccuracies. In the hard science milieu of Working Group I especially, reviewers emphasize accuracy and objectivity in the statements the writing groups make. As we will see below, comments often coalesce around the creation of or deletion of tables and graphs. I interpret this to be gatekeeping of the assessment report's anchoring devices (van der Sluijs et al. 1998)—those quantitative bits of information that hold together various social worlds within the IPCC assessment community. Since these anchoring devices form the social glue of the assessment report, the publishers decide that those most important bits of information should be printed not in text but in bold, colorful boxes with easily consumable numerical data front and center. Privileging these bits of data and the ways in which they are organized marks some of the social negotiations over the value of certain pieces of information.

A prevailing myth in science and technology implies that the more research one does, the more accurate the results. That is, more research translates into increased certainty in climate change. Accuracy is not an automatic product of an industrious hive mind but something that has to be particularly sculpted (MacKenzie 1990). Because of this, contemporary scientific assessments turn to accountability measures to ensure accuracy in both content and procedure.

According to Theodore Porter (1995), who analyzes the production of what he calls "social numbers," numbers represent objectivity, seeming to achieve a patina of neutrality that text, with its word choice, caveats, and gray areas, cannot. Both of the case studies below analyze the use of numbers in the IPCC assessment report, demonstrating how they are not only value laden but also socially negotiated both within the writing group and in the larger review process.

By writing a history of how specific IPCC numbers come to be, I am also tracking the ways in which oversight in the assessment process has built upon traditional peer review practices. The reviews, both in the formal IPCC process and in the public consumption following the reports' publication, have increasingly taken up the mantle of auditing—oversight that combines attention to accuracy as well as to accountability (Strathern 2000).

Accounting has been written about as window dressing, a tacked-on practice external to the core workings of an organization (Carruthers 1995). In contrast, some people claim that accounting is everything, playing into its ambiguity, with implications of careful bean counting as well as judgment. Accounting is a set of practices that are meant to enable decision making. As Maurer writes, "The facts of accounting are special facts: they are supposed to help people make good decisions about the management of their assets" (2002: 650). Facts emerging from accounting practices are meant to make clear the state of what is being accounted in numerical terms as well as the processes that occurred to arrive at these terms. In this sense the IPCC reviewers I discuss below are working as accountants, reviewing the draft assessment reports for information that is accurate numerically but also high in quality in terms of procedure and pedigree.

EXAMPLE 1: HIMALAYAN GLACIER TYPO

The Himalayan glaciers are undoubtedly melting: local people and scientists are observing these rapid changes in real time (Tandong et al. 2012). The rate of change is so dramatic that the authors of the chapter on Asia in Working Group II's assessment report selected the Himalayan glaciers as one of their two focused case studies.[1] In this example, I look at how a particular paragraph emerged and then came under scrutiny—both during the review process and after publication. One sentence in this paragraph contained a typo that triggered, along with the hacked email controversy Climategate, an external review of the IPCC. How did such a typo persist under a telescoping peer review process? By peeling away the layers of peer review from the published sentence, we learn how a particular error can gain epistemic legitimacy.

The First Order Draft stated, "Glaciers in the Himalaya are receding faster than in any other part of the world (see Table 10.12 below) and, if the present rate continues, the likelihood of them disappearing by the year 2035 and perhaps sooner is very high if the Earth keeps getting warmer at the current rate. The glaciers will be decaying at rapid, catastrophic rates. Its total area will shrink from the present 500,000 to 100,000 km² by the year 2035" (IPCC 2006: 54). The only expert comments from the First Order Draft were requests to remove the table; no reasons for doing so were given. These comments were deemed by the chapter authors to be "irrelevant editorial comments."

The Second Order Draft text is identical, though the accompanying table was reformatted and renumbered to account for deleted tables in other sections. There are two sets of review comments for the Second Order Draft: another expert round and a governmental round. All experts who reviewed the First Order Draft are invited to review the next draft, and other experts, users, stakeholders, and government representatives are also in the reviewing mix.

In the expert round, two people made comments on this paragraph: Poh Poh Wong of the National University of Singapore and David Saltz of Ben Gurion University. Wong, a geographer, asked for clarification on the table corresponding to the paragraph so that the relatively dramatic retreat rate of the Himalayan glaciers would stand out in comparison to glaciers in other locations. Saltz, an American-born desert ecologist working in Israel, conversely, poked at the content of the sentence itself. He asked what "its" refers to, and got a one-word response: "glaciers," with no changes made to the text. He also pointed out the inconsistency between a glacier 100,000 km² in size and one that has "disappeared." The writing team's comment, "missed to clarify this one," was appropriate, since they did not clarify this in the final draft either. Also, Clair Hanson, representing the IPCC's Technical Support Unit, noted that there is "only one reference in this whole section" on Himalayan glaciers. There are no references in the paragraph of interest here. The writing team wrote "more references added" as their response.

In the governmental comments, the government of Japan took the section to task as well, also noting the lack of credible references. So-called governments in the IPCC are state delegations, usually made up of representatives from relevant state agencies for international diplomacy, scientific research, and development. Sometimes, governments invite scientists to sit on their delegations.

The government of Japan found the projection of 2035 so alarming that they thought it should be moved to the Summary for Policymakers, a move that would give the statement more political visibility. The comment also asked the writing team to clarify the likelihood of this projection, referring to a box that standardizes phrases like "high" and "very high" into numerical translations of certainty. The Japanese governmental reviewer wrote, "This seems to be a very important statement, possibly should be in the SPM [Summary for Policymakers], but is buried in the middle of this chapter. What is the confidence level/certainty? (i.e. 'the likelihood of the glaciers disappearing is very high' is at which level of likelihood? (ref. to Box TS-1, 'Description of Likelihood'). Also in this paragraph, the use of 'will' is ambiguous and should be replaced with appropriate likelihood/confidence level terminology." The writing team responded, "Appropriate revisions and editing made."

As a comment on the entire chapter, the Japanese government also raised a concern about its reliance on gray literature—literature produced by NGOs or other assessments—instead of on peer-reviewed, published research. The commenter wrote, "There is a general concern with the lack of references throughout the Asia chapter. Because this is supposed to be an assessment of research since TAR [the Third Assessment Report, published by the IPCC six years previously], the reader wonders when reading sections where there are no references just exactly where the information is coming from. Moreover, lack of references seriously jeopardizes the scientific validity of this chapter on Asia. Another concern is a heavy use of FAO, WHO, UN-HABITAT, among other international organizations' material for references." The writing team responded by noting, "More references were added."

In the final, published text, the paragraph had some slight edits from the one in the Second Order Draft. It read: "Glaciers in the Himalaya are receding faster than in any other part of the world (see Table 10.9) and, if the present rate continues, the likelihood of them disappearing by the year 2035 and perhaps sooner is very high if the Earth keeps warming at the current rate. Its total area will likely shrink from the present 500,000 to 100,000 km^2 by the year 2035 (WWF, 2005)." One sentence had been removed on the advice of Saltz, and the wording had some minor changes. The most significant edit, however, is the addition of a citation of the World Wide Fund for Nature (WWF) at the end of the paragraph.

This citation refers not to research conducted by the WWF but to a piece of gray literature. The WWF's source was a report from 1999 of the International Commission for Snow and Ice, which the NGO quotes as stating, "Glaciers in the Himalayas are receding faster than in any other part of the world and the livelihood [*sic*] of them disappearing by the year 2035 is very high" (WWF 2005:29). But this is not what the International Commission for Snow and Ice's report actually said. Rather, the commission's report cited a UNESCO study of 1996 which stated that "the extrapolar glaciation of the Earth will be decaying at rapid, catastrophic rates—its total area will shrink from 500,000 to 100,000 km^2; by the year *2350* [emphasis added]."[2] Thus there was an erroneous transposition of the time period in which this dramatic change will take place, from 2350 in the original to 2035 in subsequent citations (Kotlaykov 1996). In addition, there was an erroneous transposition in the geographic scope, from reference to the ice coverage of all glaciers outside of Antarctica to a reference to Himalayan glaciers only.

It is difficult to trace how the errors were identified: they exploded in the skeptic blogosphere around December 1, 2009, in the aftermath of the Climategate email hacks and the publication of an Indian report on Himalayan glaciers which disagreed with the rapid melt prediction stated by the IPCC (Khandekar 2009; Mahony 2014b; Raina 2009). When these errors became public, a huge controversy ensued. The dispute was exacerbated by the finding of the climate oversight NGO Climate Science Watch that much of the text of the WWF report was in fact plagiarized from an article in 1999 titled "Glaciers Beating Retreat," published unattributed in the environmental activist magazine *Down to Earth*. In response to this controversy, the WWF report now contains a corrigendum, which reads, "This statement was used in good faith but it is now clear that this was erroneous and should be disregarded" (WWF 2005: i).

At the simplest level, the year 2035 was a transposition of numbers: the correct number could be traced to 2350, a more realistic projection in light of current research. This mistake can be described as the mere repetition of a singular typing error. However, in the political context of the IPCC, the mistake was amplified to pick away at the legitimacy of the organization. This error raises questions about the review process of non-IPCC documents, such as the World Wide Fund for Nature and International Commission on Snow and Ice reports. Most important, it shows a flaw in the review process of the IPCC itself, as a mode both of ensuring accuracy and of maintaining a high quality of sources.

EXAMPLE 2: ICE SHEETS IN THE FOURTH ASSESSMENT REPORT

Working Group I also had a contentious issue regarding glaciers in the Fourth Assessment Report. In this example, the scope is much larger, and the drama plays out all the way through the Summary for Policy Makers. In this case, the assessors are concerned with projecting the rapid disintegration of ice sheets: huge land-based glacial systems. The melting of these ice sheets, which are located in Greenland and Antarctica, has the potential to significantly—even disastrously—raise global sea level. While the Himalayan glacier number is clearly a typo, most of the assessors involved in this ice sheet example stand by their numbers as representing the best available knowledge in peer-reviewed literature at the time of the IPCC deadline. The public debate following the publication of these numbers highlighted the vast uncertainty surrounding the rapid disintegration of polar ice sheets as well as a conservatism among climate scientists (Brysse et al. 2012).

How did previous assessment reports handle sea level rise projections in relationship to polar ice sheets? Assessors have traditionally struggled to incorporate what are called nonlinear events into more linear patterns of climate change such as ocean acidification and atmospheric warming. The Greenland and Antarctic Ice Sheets contribute to sea level rise linearly through the mechanisms of accumulating mass (adding ice to the ice sheets, which results in negative sea level rise) or melting (which increases sea level rise). These contributions are usually small on a global scale, and measuring and predicting them is somewhat more straightforward. What is more concerning is the possibility of the nonlinear event which glaciologists call rapid disintegration, that is, the collapse of the ice sheet. Such an event would be relatively rapid (one hundred to three hundred years) and would raise global sea levels to a crisis state for human settlements. The West Antarctic Ice Sheet alone is projected to raise global sea level by 3.3–6m if it disintegrates (Alley and Whillans 1991; Bamber et al. 2009; Lythe et al. 2001).

In the IPCC's Third Assessment Report (2001), total sea level rise projections through the year 2100—0.09–0.88m—decreased slightly from those in the Second Assessment Report (which were 0.13–0.94m). The assessors wrote in the Summary for Policymakers that "despite the higher temperature change projections in this assessment, the sea level projections are slightly lower, primarily due to the use of improved models, which give a smaller contribution from glaciers and ice sheets" (2001: 16).

The Fourth Assessment Report sea level rise projections offered an even lower number, marking the third consecutive tuning down of sea level rise estimates. The numerical projections are most visible in the table that was presented in both chapter 10 and the Summary for Policymakers (table 4.1). Taken together, the various scenarios in this chart result in a range in sea level rise of 0.18–0.59m. Unlike the Third Assessment Report, this lower upper bound is not due to improved modeling. Instead, it reflects the fact that the assessors decided the science was too uncertain to incorporate the projections of multicentury changes from the West Antarctic Ice Sheet that had been included in the former report. The assessors note this exclusion in

Table 4.1. Table SPM.3. Projected global average surface warming and sea level rise at the end of the 21st century. (IPCC 2007)

	Temperature change (°C at 2090–2099 relative to 1980–1999)[a]		Sea Level Rise (m at 2090–2099 relative to 1980–1999)
Case	Best estimate	*Likely* range	Model-based range excluding future rapid dynamical changes in ice flow
Constant Year 2000 concentrations[b]	0.6	0.3–0.9	NA
B1 scenario	1.8	1.1–2.9	0.18–0.38
A1T scenario	2.4	1.4–3.8	0.20–0.45
B2 scenario	2.4	1.4–3.8	0.20–0.43
A1B scenario	2.8	1.7–4.4	0.21–0.48
A2 scenario	3.4	2.0–5.4	0.23–0.51
A1FI scenario	4.0	2.4–6.4	0.26–0.59

Table notes:

[a] These estimates are assessed from a hierarchy of models that encompass a simple climate model, several Earth System Models of Intermediate Complexity and a large number of Atmosphere–Ocean General Circulation Models (AOGCMs).

[b] Year 2000 constant composition is derived from AOGCMs only.

the caveat in the table that reads, "Model-based range excluding future rapid dynamical changes in ice flow." To translate: the assessors left the incremental contributions of accumulation and melting from the ice sheets in their projection but left out any projections of rapid disintegration. How was this projection treated in the review drafts of the Summary for Policymakers?

In the Second Order Draft (4/4/2006) of the summary, reviewed by experts as well as governments, there was no chart like the one shown in table 4.1. Instead, the authors explained their decision in the text, writing as follows:

- Sea level rise commitments have much longer timescales than warming commitments, owing to slow processes that mix heat into the deep ocean. By 2100, sea level rise is projected to range from 0.14–0.43m for the A1B scenario concentrations for example, but would be expected to show larger increases (0.3–0.8m due to thermal expansion) in the next two centuries even after stabilization at those concentrations [10.6, 10.7]
- Changes in the ice sheets could significantly affect future sea level rise commitment. Models suggest that a global average warming of 3°C above present levels would cause widespread mass loss of the Greenland Ice Sheet if sustained for several centuries, initially contributing up to 0.4m sea level rise per century. The melting rate would increase if dynamical processes increase the rate of ice flow, as suggested by some recent observations. This level of warming could occur during the 21st century depending upon the climate sensitivity and the emission scenario, and is comparable to that of the last interglacial period about 125,000 years ago, when paleoclimate data suggest that widespread Arctic melting contributed several meters of sea level rise [10.7]
- The Antarctic Ice Sheet is projected to behave differently from that of Greenland because it is too cold for widespread surface melting. It is expected to gain ice through increased snowfall in the 21st century, acting to reduce global sea level rise by about 0.1m per century. However, in response to weakening of ice shelves by ocean warming or surface melting at the margins, ice flow could accelerate. Such effects could offset or outweigh increased snowfall but are uncertain. [10.7]

In the archived comments, some reviewers were perturbed by the low projections, which were even lower in this earlier draft than they were in the final report. Others asked authors to clarify what was meant by "dynamical processes," which is a technical, accurate scientific term but one that has little meaning for the nonexperts for whom the summary was written. Other reviewers addressed the deep uncertainty reflected by the West Antarctic Ice Sheet models, which had been considered as the gold standard, given that these models have not been able to capture the rapid changes observed on the ice sheet over the past ten years.

John Hunter, an oceanographer at the Antarctic Climate and Ecosystems Cooperative Research Centre at the University of Tasmania in Australia, wrote, in response to the bullet point (or "dot point") draft text above, "This summary of sea-level rise projections is confusing in the extreme. It is not clear whether the contributions for the three dot points should be added. It is also evident that a full range of SRES scenarios has not been covered.[3] How does the lay reader know what the 'A1B Scenario' is? Nowhere in the SPM [Summary for Policymakers] is the A1B Scenario described. These dot points (and other parts of the SPM, if they are as bad as this) are the laying of a political minefield, with projections open to whatever interpretations the quoter desires. We need to know the full range of projections for the full range of model uncertainties (SPM-1395)." Hunter's concern over the clear depiction of science showed an abiding understanding of the political power of the IPCC reports, the scrutiny they undergo, and potential misinterpretations.

Michael MacCracken, a former U.S. government climate scientist who is now the chief scientist of climate change programs at the NGO the Climate Institute, worded his comments even more strongly, confident that the numbers were wrong:

> This estimate of sea level rise is much too small—it is simply not credible given what occurred in the twentieth century and the acceleration evident in the last decade. This range indicates that the rise over the twenty-first century could be about 25 percent less than observed for the twentieth century—and there is no way that build-up on Antarctica could accomplish this in the face of the increases that are well-established to occur from mountain glaciers and the accelerating rate of change—now up to a rate of .3m per century. Given how Greenland deteriorations [are] starting, the notion that .43m is the upper limit seems patently absurd. At least, as compared to the TAR, IPCC is not saying this is 'very likely' etc.—but this range is far too low for any scenario—and the large increases could also come from deterioration of Greenland, as paleo evidence has suggested is possible. I am just baffled by this estimate (SPM-1397).

MacCracken's intervention, among others, encouraged the assessors to take another look at their conclusions and led to substantial revisions in the Third Order Draft (10/27/2006), which underwent a governmental review. The Third Order Review text reads as follows:

- Contraction of the Greenland ice sheet is projected to continue to contribute to sea level rise after 2100. Current models suggest that a global average warming (relative to pre-industrial values) of 1.9 to 4.6°C would lead to virtually complete elimination of the Greenland ice sheet and a resulting sea level rise of about 7m, if sustained for millennia. These temperatures are comparable to those inferred for the last interglacial period 125,000 years ago, when paleoclimatic information suggests reductions of polar ice extent and 4 to 6m of sea level rise. {6.4, 10.7}
- Dynamical processes not included in current models but suggested by recent observations could increase the vulnerability of the ice sheets to warming, increasing future sea level rise. Understanding of these processes is limited and there is no consensus on their magnitude. {4.6, 10.7}
- Current global model studies project that the Antarctic ice sheet will remain too cold for widespread surface melting and is expected to gain in mass due to increased snowfall. However, net loss of ice mass could occur if dynamical ice discharge dominates the ice sheet mass balance. {10.7}
- 21st century anthropogenic carbon dioxide emissions will contribute to warming and sea level rise for more than a millennium, due to the timescales required for removal of this gas. {7.3, 10.3}

There were many comments from governmental representatives on these bullet points. Governments from Denmark, the European Community, Germany, the Netherlands, New Zealand, Sweden, the United Kingdom, and the United States made strong comments about their confusion over the "dynamical processes" language. The governments of Belgium, Germany, Italy, New Zealand, the United Kingdom, and the United States also commented on the problems of credibly modeling ice sheets.

With the input of peer expert and governmental reviewers, the final set of sea level projection numbers and the design of the SPM.3 chart (see table 4.1) were negotiated at the Summary for Policy Makers meeting in France with the Coordinating Lead Authors and national delegations. They decided that their understanding of rapid dynamical ice flow was too uncertain to represent in the sea level rise numbers at all, though there is a calculation in the table to account for the steady melt rate (which is very low), and there is text that talks about how the potential sea level rise contribution of ice sheet collapse could be very high.

It is notable, however, that the numbers in the table are not derived purely from peer-reviewed published material. They are the result of a number of informal calculations conducted by a handful of assessors who were attempting to bridge some of the gaps in published material. These numbers were not made official until after the last round of commenting was completed. In the process, the assessors rejected at least one other informal set of numbers, a calculation put together by Stefan Rahmstorf, who eventually published his numbers in *Science*, a journal that some may see as the gold standard of scientific peer review (2007, recorded phone interview between Rahmstorf and O'Reilly, August 30, 2010). I have written up the story of this negotiation elsewhere (O'Reilly et al. 2012): what I am interested in for the purposes of this chapter is the treatment of the numbers.

In the case of the sea level rise figures, which excluded the potential contribution of ice sheet melting, the table served as an anchoring device, drawing attention away from the caveats in the text (Oppenheimer et al. 2007). The figures in question are also produced through the IPCC process. They are not sloppy typos like the Himalayan glacier number but politically loaded numbers. Some scientists (including some of the authors) think they are the best representation of the state of the science at that moment. Others think they are an underrepresentation, a misrepresentation, or a flat-out mistake.

HYBRID, CASUAL, "MUTT" NUMBERS

How and when do IPCC gatekeeping practices such as governmental and peer review, gap filling, and new knowledge allow casual numbers to filter into a document that is expected to assess only existing climate science? Peer review is a standard means of verifying scientific information, and, as described above, the IPCC process relies on a concerted series of peer review rounds. Each of these review comments—at minimum, hundreds of them for each chapter and often thousands—has to be responded to by the authors. The reviewers have a heavy workload as well. These documents are not straightforward research articles, with apparent methods and a single point of inquiry. The amount of information being synthesized is massive, complex, interdisciplinary, and sometimes highly uncertain. The Himalayan glacier typo can be considered a failure in the peer-review process. The typo was discovered through an informal audit and gained traction through politically savvy climate skeptics. In the case of the West Antarctic Ice Sheet number, on the other hand, there was not a failure in peer review—it worked as expected.

Yet both of the numbers above are casual numbers, mutts. IPCC authors expect their numbers to have a pedigree, a trail of paperwork that neatly and clearly traces a genealogy of the knowledge presented and assessed. This genealogy is created through citation practices, the referencing of information that is being discussed and the information that forms the base for the new directions one's work has taken. In the case of the Himalayan glacier typo, the genealogy fails. The reviewers urge the authors to include a citation, but the citation ultimately provided is a weak one. It is not from a peer-reviewed science publication but from an NGO. It is a gray source, an assessment, so the citation is an assessment citing an assessment: hybrid vigor does not stand up here. By extension, the casual number produced in the sea level rise table was previously unpublished, not a peer-reviewed number calculated by the writing team. This casual number is a new number and represents new knowledge, but its pedigree is muddled and murky.

In the IPCC, we see scientists, authors, and reviewers moving from a classic scientific peer-review process to an audit process. Both skeptics and IPCC authors are shifting toward audit practices. While scientific peer review evaluates ideas, methods, and findings for quality and accuracy, audits call up accountability, both in a numerical sense, in which the accountant makes numbers comprehensible, and in an ethical sense, where people's activities, according to Marilyn Strathern (2000), are subject to oversight and review. Auditing can be seen as a science itself, and so auditing science, a conceptual stretch that seems to be gaining momentum, is starting to gain formal traction among IPCC authors and bureaucrats. The casual numbers analyzed above are typically identified through independent or self-auditing of the IPCC. The use of casual numbers—controversial outside the writing groups but considered conventional, unremarkable, and necessary within them—does not stand up to standardized institutional review.

In conclusion, convincing people to worry about—and ultimately act to reverse—climate change requires a flexible set of strategies. One prevalent strategy is the calling up of charismatic data—real-time, dramatic, breathtaking events—to make climate change appear palpable and urgent. Another is the cultivation of scientific assessments, most clearly epitomized by the IPCC. In IPCC assessment reports, charismatic climate events are subsumed while the standardization of various sources, methods, and findings—all peer-reviewed scientific research—is emphasized. This, too, is convincing work, taking place under the mantle of scientific objectivity. In this chapter, I analyzed two examples in which the most recent IPCC assessment report could be considered to fail: the first a typo, the second a consensus agreement to omit a significant component of a glacial system.

Projections of future climate change cannot be anything but guesswork—highly technical, expensive, educated guesswork—but always with an element of "gap filling," prognostication, or intuition. How this guesswork is performed matters—and how it is represented by scientists matters as well. While back-of-the-envelope calculations are legitimate work—indeed, such calculations may be said to be a large part of what scientific reasoning is—such informal practices seem mysterious or even fraudulent to nonexpert observers. Instead of glossing over the controversies and disagreements implicit in the production of IPCC assessment reports, I work with participating scientists to analyze how their micropractices inform the state of climate science knowledge. By bringing this guesswork under critical scrutiny, we can understand how these micropractices inform major scientific projections and start to plan for a changing environment.

NOTES

I would like to acknowledge the collegial input of this book's editors and coauthors, as well as the mentorship of Naomi Oreskes and Michael Oppenheimer.

1. The other case study examined the megadeltas of Asian rivers and did not come under the scrutiny that the Himalayan glacier case study received.
2. The 1999 International Commission on Snow and Ice report and the 1996 report that it cites have now been removed from their online repositories. I continue to look for the originals but am citing directly from a blog post on climatesciencewatch.org, whose investigation into the Himalayan glacier typo paralleled my own.
3. SRES scenarios refer to the Special Report on Emissions Scenarios published by the IPCC that provided multiple future emissions scenarios for international climate modelers to input and run, so that the data could be compared between models. The A1B is one such scenario that is characterized by a vision of a future with a "more integrated world," with a population reaching nine billion around 2050 and then gradually declining, rapid economic growth, and "a balanced emphasis on all fuel sources," both carbon and noncarbon based (IPCC 2000).

REFERENCES CITED

Alley, R. B., and I. M. Whillans. 1991. "Changes in the West Antarctic Ice Sheet." *Science* 254, no. 5034: 959–63.

Bamber, J. L., R. E. M. Riva, B. L. A. Vermeersen, and A. M. LeBroq. 2009. "Reassessment of the Potential Sea-Level Rise from a Collapse of the West Antarctic Ice Sheet." *Science* 324, no. 5929: 901–3.

Brysse, K., N. Oreskes, J. O'Reilly, and M. Oppenheimer. 2012. "Climate Change Predictions: Erring on the Side of Least Drama?" *Global Environmental Change* 23: 327–37.

Carruthers, B. 1995. "Accounting, Ambiguity, and the New Institutionalism." *Accounting, Organisations, and Society* 20: 313–28.

Choy, T. 2011. *Ecologies of Comparison: An Ethnography of Endangerment in Hong Kong.* Durham: Duke University Press.

Down to Earth. 1999. "Glaciers Beating Retreat." Last modified April 30, 1999. http://www.downtoearth.org.in/content/glaciers-beating-retreat.

Ferguson, J. 1990. *The Anti-Politics Machine: 'Development,' Depoliticization, and Bureaucratic Power in Lesotho.* New York: Cambridge University Press.

InterAcademy Council (IAC). 2010. *Committee to Review the Intergovernmental Panel on Climate Change. Climate Change Assessments: Review of the Processes and Procedures of the IPCC.* The Netherlands: Bejo Druk and Print.

Intergovernmental Panel on Climate Change (IPCC). 2000. *Special Report on Emissions Scenarios: A Special Report of Working Group III of the Intergovernmental Panel on Climate Change,* Edited by N. Nakićenović and R. Swart. Cambridge: Cambridge University Press.

———. 2001. *Climate Change 2001: The Scientific Basis. Contribution of Working Group I to the Third Assessment Report of the Intergovernmental Panel on Climate Change.* Edited by J. T. Houghton, Y. Ding, D. J. Griggs, M. Noguer, P. J. van der Linden, X. Dai, K. Maskell, and C. A. Johnson. New York: Cambridge University Press.

———. 2006. *Working Group II Drafts and Review Comments for the Fourth Assessment Report.* Accessed October 4, 2013. http://www.ipcc-wg2.gov/publications/AR4/ar4review.htm.

———. 2007a. *Climate Change 2007: The Physical Science Basis. Contribution of Working Group I to the Fourth Assessment Report of the Intergovernmental Panel on Climate Change.* Edited by S. Solomon, D. Qin, M. Manning, Z. Chen, M. Marquis, K. B. Averyt, et al. New York: Cambridge University Press.

———. 2007b. *Climate Change 2007: Impacts, Adaptation, and Vulnerability. Contribution of Working Group II to the Fourth Assessment Report of the Intergovernmental Panel on Climate Change.* Edited by M. L. Parry, O. F. Canziani, J. P. Palutikof, P. J. van der Linden, and C. E. Hanson. New York: Cambridge University Press.

International Commission for Snow and Ice (ICSI). 1999. *A Report by the Working Group on Himalayan Glaciology (WGHG) of the International Commission for Snow and Ice (ICSI).*

Jasanoff, S. 1990. *The Fifth Branch: Science Advisors as Policymakers.* Cambridge: Harvard University Press.

Khandekar, M. L. 2009. "Global warming and the glacier melt-down debate: a tempest in a teapot?" Accessed May 6, 2013. Blog post http://pielkeclimatesci.wordpress.com/2009/12/01/global-warming-and-glacier-melt-down-debate-a-tempest-in-a-teapot/.

Kotlaykov, V. M. 1996. *Variations of Snow and Ice in the Past and at Present on a Global and Regional Scale.* Paris: UNESCO.

Lamont, M. 2009. *How Professors Think: Inside the Curious World of Academic Judgment.* Cambridge: Harvard University Press.

Latour, B. 1990. "Drawing Things Together." In *Representation in Scientific Practice,* ed. M. Lynch and S. Woolgar, 19–68. Cambridge: MIT Press.

Lythe, M. B., D. G. Vaughan, and BEDMAP Consortium. 2001. "BEDMAP: A New Ice Thickness and Subglacial Topographic Model of Antarctica." *Journal of Geophysical Research* 106, no. B6: 11335–51.

MacKenzie, D. 1990. *Inventing Accuracy: A Historical Sociology of Nuclear Missile Guidance.* Cambridge: MIT Press.

Mahony, M. 2014a. "Climate Change and the Geographies of Objectivity: the Case of the IPCC's Burning Embers Diagram." *Transactions of the Institute of British Geographers. Doi 10:1111.*

———. 2014b. "The Predictive State: Science, Territory and the Future of the Indian Climate." *Social Studies of Science* 44, no. 1: 109–33.

Maurer, B. 2002. "Anthropological and Accounting Knowledge in Islamic Banking and Finance: Rethinking Critical Accounts." *Journal of the Royal Anthropological Institute* 8, no. 4: 645–67.

Oppenheimer, M., B. C. O'Neill, M. Webster, and S. Agrawala. 2007. "Climate Change: The Limits of Consensus." *Science* 317, no. 5844: 1505–6.

O'Reilly, J., N. Oreskes, and M. Oppenheimer. 2012. "The Rapid Disintegration of Projections: The West Antarctic Ice Sheet and the Intergovernmental Panel on Climate Change." *Social Studies of Science* 42, no. 5: 709–31.

Porter, T. M. 1995. *Trust in Numbers: The Pursuit of Objectivity in Science and Public Life.* Princeton: Princeton University Press.

Rahmstorf, S. 2007. "A Semi-Empirical Approach to Projecting Future Sea-Level Rise." *Science* 315: 368–70.

Raina, V. K. 2009. "Himalayan Glaciers: A State-of-the-Art Review of Glacial Studies, Glacial Retreat, and Climate Change." *Science and Public Policy Institute: Ministry of Environment and Forests, Government of India and G. B. Pant Institute of Himalayan Environment and Development.*

Riles, A. 1998. "Infinity Within the Brackets." *American Ethnologist* 25, no. 3: 378–98.

Shapin, S. 1995. *The Social History of Truth: Civility and Science in Seventeenth-Century England.* Chicago: University of Chicago Press.

Strathern, M., ed. 2000. *Audit Cultures: Anthropological Studies in Accountability, Ethics and the Academy.* New York: Routledge.

Tandong, Y., L. Thompson, W. Yang, W. Yu, Y. Gao, X. Guo, X. Yang, K. Duan, H. Zhao, B. Xu, J. Pu, A. Lu, Y. Xiang, D. B. Kattel, and D. Joswiak. 2012. "Different

Glacier Status with Atmospheric Circulations in Tibetan Plateau and Surroundings." *Nature Climate Change* 2: 663–67.

van der Sluijs, J., J. van Eijndhoven, S. Shackley, and B. Wynne. 1998. "Anchoring Devices in Science for Policy: The Case of Consensus Around Climate Sensitivity." *Social Studies of Science* 28, no. 2: 291–323.

World Wide Fund for Nature (WWF). 2005. "Glaciers, Glacier Retreat, and the Subsequent Impacts in Nepal, India, and China." WWF Nepal Program.

Chapter 5 **Scale and Agency: Climate Change and the Future of Egypt's Water**

Jessica Barnes

In November 2011, ten days before the Durban climate negotiations began, the British *Guardian* newspaper ran a story titled, "Drier, Hotter: Can Egypt Escape its Climate Future?" The article outlined three main dimensions to Egypt's climate future: "inexorably rising" sea levels, "unbearably hot" temperatures, and uncertainty about the country's water supply, which could increase by 30 percent but could, alternatively, decline by 70 percent. Under an image of a pyramid against Cairo's skyline, the article warned of "a perfect storm brewing" (Vidal 2011).

While rising sea levels and hotter temperatures undoubtedly pose serious problems for Egypt, the third ingredient of this perfect storm—the possibility of a dramatic shift in the water supply—is perhaps the most profound. As the newspaper article notes, whether or not such a shift in water availability will occur is uncertain. The future of Egypt's water supply lies in the future of the Nile River. Rising in the moister climates of the East African Highlands, the Nile provides 96 percent of Egypt's water, feeding the country's fields (where 90 percent of the water is consumed), cities, and industries (fig. 5.1). Owing to the

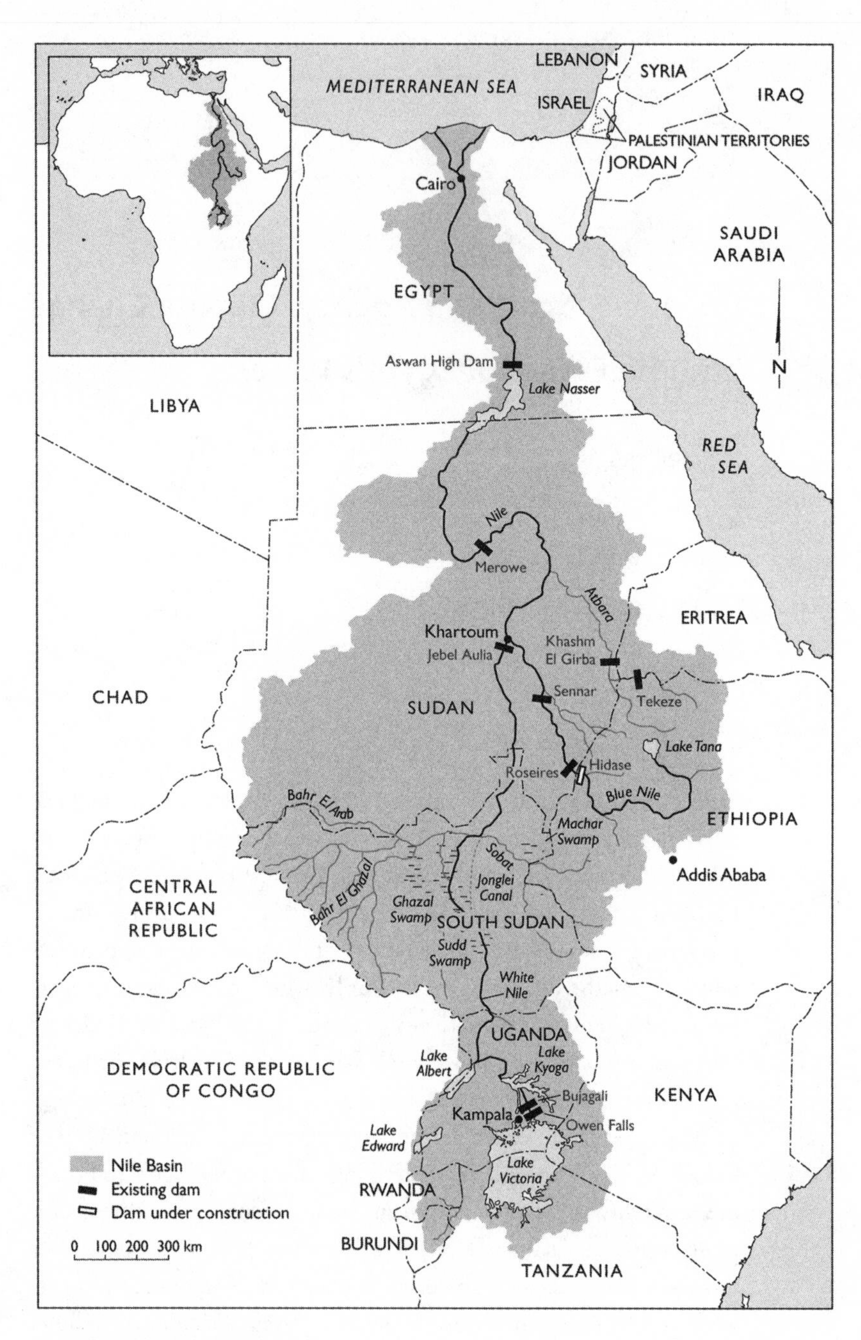

5.1: The Nile Basin. Bill Nelson.

complexity of precipitation dynamics, it is impossible to predict with confidence what will happen to rainfall in the river's source regions under climate change in coming decades. Rainfall may increase, leading to increased river flows, or decrease, leading to reduced flows. Either way, the impact on Egypt will be far-reaching.

Given the combination of both an uncertain outlook and the potential for a significant shift in river flow patterns, climate change has become a prominent concern in discussions among water experts working in Egypt and the Nile Basin. Recent years have seen a flourishing of internationally funded initiatives, workshops, and panel discussions on climate change impacts on the basin. Water specialists have urged governments to introduce adaptation strategies to plan for the changes in water supply that they may see in coming years. Donor resources have poured into projects to improve the basin countries' capacities to respond to this challenge.

Yet climate change is not the only factor that will shape water availability in Egypt in coming years. Through an analysis of some of the contrasting ways of looking at Egypt's water futures, this chapter probes how variously positioned actors attach different degrees of weight to climate change as an explanatory variable. In doing so, the chapter reveals how a relationship that seems so clearly defined at a large scale—the fact that changing concentrations of greenhouse gases are affecting patterns of temperature and precipitation and therefore water availability—becomes much more slippery when examined in a particular context. In addition, it raises questions about the notions of agency implicit in discussions of climate change impacts. Defined by Ahearn (2001: 112) as "the socioculturally mediated capacity to act," agency has long been of interest among anthropologists and social scientists. The case of anthropogenic climate change, which comprises both physical dynamics of ocean–atmosphere interaction and societal drivers of atmospheric composition and land use, calls on us to think about the intersections of natural and human agency in shaping environmental outcomes. In comparing how a range of actors, including scientists, government officials, international experts, irrigation engineers, and farmers, understand the future of a climate-linked resource, we see how the assignment of agency, the pinpointing of where power over development trajectories lies, is both subjective and political (see also Dove et al. 2006; Jonsdottir 2013).[1]

This chapter examines three visions of Egypt's water future. In each case, the relative scarcity or abundance of water in Egypt is a key concern, but the factors perceived to drive that availability differ. I start with the international water conferences where climate change is a central topic on the agenda. Drawing on ethnographic data from the Sixth World Water Forum, held in Marseille in March 2012, and the Second Arab Water Forum, held in Cairo in November 2011, I look first at how scientists, government officials, and international experts who move within these transnational circles understand the risks posed to Egypt by climate change. I explore how they see water futures through a lens of models, which capture water in a series of equations that reproduce the hydrological fluxes of precipitation, runoff, and evapotranspiration over the scale of the river basin. I show how these discussions sideline questions of human agency, paying little attention to the political decisions about water allocation and access that will ultimately shape how changes in Nile flows are experienced within Egypt.

In the second section I look at another set of conversations around Egypt's water futures. These conversations also focus on the scale of the Nile Basin and Egypt's position within it. They too take place at international water conferences as well as behind the closed doors of political negotiations, and involve scientists, government officials, and international experts. Yet their focus is different. Rather than looking at the future of Egypt's water supply in terms of changing precipitation patterns, these discussions emphasize the actions of countries that lie upstream within the Nile Basin and how they will affect Nile flows to Egypt in coming years. Many of those participating in these debates are well aware that climate change is taking place and that precipitation patterns are shifting. Indeed, in other contexts they also take part in conversations about climate change and efforts to model its impacts. But their primary concern lies with what they perceive to be a more immediate threat: the construction of upstream dams along the Nile. They weigh geopolitics over climate-discharge relations, human agency over natural agency.

Third, I examine the site where most of the water in question is actually being used: the fields and canals of rural Egypt. Over the course of a year's fieldwork that I conducted in the predominantly agricultural province of Fayoum in 2007–8, I never heard any discussion among either farmers or irrigation engineers about climate change or its potential impact on water resources. Water variability—the presence or absence of water in the canals—is a matter of constant concern. Farmers and engineers see this variability, though, not in terms of climatic parameters operating on the scale of the river basin as a whole, but in terms of more localized structures of water control. They envisage water futures as lying in the hands of those who operate these dams and weirs.

The future is therefore imagined through various scales, which are underpinned by contrasting notions of natural and human agency.[2] As the lens shifts from a larger scale—that of the river basin—to a smaller scale—that of an irrigation canal—the import of climate as a driving force recedes and other, more overtly political factors come into focus. As future water availability moves from the abstract (a figure of the billion cubic meters that will flow across Egypt's border in a year) to the immediate (a question of whether there will be enough water in a canal for a farmer to grow a crop next year), it becomes more embedded in specific political relationships and less amenable to the types of analysis and answers provided by climate scientists.

In juxtaposing these diverse ways of looking at Egypt's water futures, the chapter illustrates both the power and the limits of what Mike Hulme (2011) terms "climate reductionism." Climate reductionism constitutes a way of seeing the world in which the future of societies and environments is seen in terms of climate alone. This reductionism, Hulme argues, has become a sign of our times as increasing concern about anthropogenic climate change has led to disproportionate attention being placed on climate over other factors that shape societies and their interactions with the physical world. Hulme sees climate models, which are unable to incorporate social, economic, political, and cultural variables in anything more than crude terms, but whose predictions are becoming increasingly influential, as being pivotal to this kind of reductionism. Indeed the first vision of Egypt's water future, shaped by models of precipitation and hydrology, is illustrative of this kind of reductionism. The second and third outlooks for Egypt's water resources reveal what cannot be captured within a future written in mathematical equations and computer code. In the distinctions between these visions we see the contrast between what Shelia Jasanoff describes as the "impersonal, apolitical and universal imaginary of climate change projected by science" and the "subjective, situated and normative imaginations of human actors engaging with nature" (2010: 233).

MODEL FUTURES

As demonstrators filled Cairo's Tahrir Square in the days prior to Egypt's elections in late 2011, scientists, government officials, and international experts from the Middle East and beyond gathered at a five-star hotel on the outskirts of the city for the Second Arab Water Forum. The forum was organized by the Arab Water Council, a nonprofit regional organization closely affiliated with the Egyptian Ministry of Water Resources and Irrigation (hereafter "the ministry").[3] Climate change or, more precisely, "the risk of climate change on water: between uncertainty and cost of adaptation," was one of the forum's three central themes.

On the hotel's lower level, out of sight of the gardens where fountains spurted water into a swimming pool alongside a pyramid-shaped fitness center, a large conference room was set up with long tables. On the stage, between tall arrangements of exotic flowers, the secretary general of the World Meteorological Organization, a European scientist, gave the keynote address for the thematic session on climate change. "I'm very happy to be here," he started, "because this is a region that is particularly vulnerable to climate change, especially in water resources." After offering an overview of his organization's role in international research on climate, weather, and water resources, he summarized the major developments in climate modeling over the past few decades. Mirroring the narrative presented by many recent academic and policy studies on climate change impacts in the Middle East (GIZ 2011; Sowers et al. 2011; Tolba and Saab 2009), his comments set the contours of the debate: the Middle East as a region in which climate change threatens to exacerbate already noteworthy water problems, and models as the tool to predict these possible futures and facilitate adaptation. This is a frame, therefore, that situates the future of the region's water resources firmly in terms of climate change rather than human decision making.

In the next panel, scientists from Egypt's water ministry presented their research on climate change impacts and Nile flows. They described how they start by taking the output from an ensemble of seventeen general circulation models to see how climatic parameters will change for their region through 2050. Then, in order to obtain predictions for temperature and precipitation over the Nile Basin at a higher spatial resolution than the 250km grid cell of most general circulation models, they use a regional climate model to downscale these model outputs. Their goal in doing so is to obtain the most accurate climate data for the Nile Basin possible, which they can then use as their predictor of water futures.

So far this group of Egyptian scientists has looked at climate predictions only under one scenario of greenhouse gas emissions (IPCC scenario A1B). This is revealing. Emissions scenarios, which relate quantities of greenhouse gases in the atmosphere to various social, economic, and political trajectories, are one of the few ways in which human agency is incorporated in a climate model. (In the case of the A1B scenario, the trajectory is one of rapid economic growth in an integrated world, with population declining after reaching a peak of nine million in 2050 and a balanced emphasis on fossil and nonfossil energy sources.) By running their models with different emissions scenarios, scientists can explore a range of social futures and thus avoid the kind of impact assessment methodology that Hulme (2011) identifies as being climate reductionist. By looking at only one emissions scenario, therefore, the Egyptian scientists reduce their outlook to a singular vision of social, cultural, and political change (see also Tsing 2005: 102–6). They are fully aware that this limits their understanding of the future. However, given that each run of the regional model through to 2050 takes three to four months to complete, during which time it has to be constantly monitored in case any glitches arise, they have not yet had the time or resources to explore multiple emissions scenarios. They therefore compromise social complexities for the sake of scientific expediency. The large screen at the front of the conference room depicted the results of this modeling work: a series of brightly colored maps showing projected changes in rainfall and temperature over the Nile Basin for the year 2050—a spectrum of colors to indicate a range of rainfall change ratios and greenish yellows to indicate warmer temperatures.

The final modeling step is to use these projected climatic changes to investigate future patterns of water availability. In this step, we see the translation from precipitation and evapotranspiration to discharge, and a reductionist framing of water futures in terms of climate alone. In the Arab forum panel, an Egyptian scientist described how they use a model suite called the Nile Forecast System to generate hydrological data using the results of the regional climate model. By inputting climate projections into the Nile Forecast System, which comprises a set of hydrological models that estimate soil moisture, hill slope runoff, river routing, and streamflow, they can produce annual hydrographs for Nile flows in future years. The modeling system can make predictions for Nile discharge at a number of points around the basin, but they focus on forecasting Nile flows into Lake Nasser, the reservoir behind the Aswan High Dam, which stretches across the river just shortly after it enters Egypt. In other words, by looking at the place where the river flows into the country, they concentrate on *national* water supply. This presents a homogenized view of Egypt's water future, erasing the acts of human agency—for example, the opening and closing of dams, digging and blocking of canals, channeling and diverting of water—that will ultimately determine where that water will flow and who will receive it.

The hydrological models calculate that by 2050 Nile flows into Egypt will be between –13 percent and +36 percent of their historical average (based on baseline data from 1950–2000). In other words, Egyptian scientists still do not know if Nile discharge will increase or decrease under climate change (Barnes forthcoming). While the chain of models they have been using has improved the preciseness of their estimates—a 49 percentage point range is markedly less than the 100 point range cited by the *Guardian* article—the bandwidth of uncertainty is still unsettling. A level of frustration was evident among the audience. After the scientist's presentation, a man put his hand up and asked, "Do you think the water of the Nile will increase or decrease?" He paused, then acknowledged, "Of course this is a big question, but what do *you* think?" This is a question that everyone wants an answer to. It is an answer, though, that these models are unable to provide. A future driven by precipitation patterns is a future that lies beyond the realm of predictability.

To scientists, this band of uncertainty is problematic, although not surprising, because it limits the degree of uptake among policy makers of their modeling results. During a discussion with a senior scientist in the ministry about the range of projected Nile flows, he commented that he cannot take this kind of data to those responsible for making policy. "They want a figure," he said. "We scientists in the ministry know uncertainty, but the policy makers don't know how to deal with ranges."[4] As we will see in the next section, though, it may be that the lack of response from government officials is due less to the nature of the climate modeling results than to the fact that they do not see water futures in terms of climate alone.

DOWNSTREAM FUTURES

The expo of the Sixth World Water Forum in Marseille took place in a vast hall decorated in a range of blue hues, with water features, potted plants, and bright lights. International organizations, NGOs, countries, and corporations showcased their water-related work in elaborate pavilions adorned with large photographs, maps, television screens, and piles of glossy publications. A vibrant buzz animated the exhibition, generated in part by the bars in many of the booths that served espresso in the morning, wine in the afternoon, and cocktails in the evening.

One afternoon during the forum, the Finnish pavilion hosted a side event about the Nile. The focus was on how to share the benefits of water use among the countries of the Blue Nile Basin, one of two main tributaries of the Nile, which contributes 85 percent of the river's discharge. People crammed into rows of white chairs arranged in front of a map that depicted the location of Finnish development projects around the world. The presenters, scientists and policy makers from the basin countries and experts who work for regional and international organizations on Nile Basin issues, sat on stools at the front, sharing a microphone so that they could be heard above the background noise of the exhibition.

Talk turned quickly to the Grand Renaissance Dam (known in Amharic as the Hidase Dam), which Ethiopia started to build across the upper reaches of the Blue Nile in April 2011 (see fig. 5.1). An official from Ethiopia's water ministry stressed that the dam was designed in the "full spirit of cooperation" and that it will benefit downstream countries. Indeed, he said, the Ethiopian prime minister has proposed that downstream countries might like to contribute to the cost of the dam, given that they will profit from the dam's storage capacity.[5] The audience was skeptical about the impact the dam will have on the river's flows. Their vision of water futures is firmly anchored in human agency, namely, that of the Ethiopian dam operators who will be able to control how much water is released to flow on down the Blue Nile. A Sudanese man asked how much control the downstream countries will have over the management of the dam. "If all the countries should contribute to the dam and share the costs," he said, "I need to know how the dam will be operated. For example, if there are seventy billion cubic meters stored in the dam, if you open the gates it will decimate the downstream countries!" Another Sudanese man commented, "There has been talk of benefits, benefits, benefits, but what of the costs of the dam?"

When the Egyptian scientist on the panel started his presentation, he began by giving an overview of the country's water resources. "I would like to share with you the water issues in Egypt," he said. "Egypt is vulnerable to climate change. Sea level is rising . . . and Egypt is dependent on the Nile for 96 percent of its water. *And* there is no green water in Egypt."[6] It was notable that he was the only presenter to begin his presentation in this way, with a focus on a single country's needs rather than the basin-wide context. This was no coincidence. By highlighting Egypt's lack of water resources other than the Nile, he drew a contrast with the other countries of the basin that receive considerable rainfall and thereby reasserted Egypt's claim to the shared water of the river. He used climate change vulnerability as a political tool to further justify Egypt's use of Nile water. As for Ethiopia's dam, he was circumspect. "We have to *prove* the benefits first," he declared, "and then we will talk about [sharing] the costs." For downstream countries, the choices made about how this dam is constructed and operated will have a critical bearing on their water supply. If the dam is used for hydropower generation alone, which is what the Ethiopian government has stated in public, it is unlikely to affect discharge patterns significantly. If, however, the dam is used for irrigation, as rumors suggest it might be, water losses through evapotranspiration could decrease the amount of water flowing on to downstream countries.

As in the first case, therefore, the future is one of uncertainty, but the root of this uncertainty lies not in the shifting dynamics of precipitation generation over the East African Highlands under climate change, but in the political choices made by Ethiopia about how to use the water stored behind its new dam. The variability and unpredictability lie not in nature but in culture. In this vision of the future, human agency dominates—an agency concentrated within the upstream states, which leaves downstream countries vulnerable and impotent.

Construction of this dam is the latest event in a long history of political contestation over the control of water within the Nile Basin, which spans eleven countries.[7] Until recently, Nile water allocation was governed by a treaty signed in 1959 between Sudan and Egypt, which granted 55.5 billion cubic meters of water a year to Egypt and 18.5 billion cubic meters to Sudan. Although the upstream countries were not party to the treaty, under its terms they were prevented from pursuing any activities that might compromise these flow volumes. Over the course of the late 1990s and 2000s, the basin countries drew up a new agreement, the Cooperative Framework Agreement (CFA). Designed to be basin-wide in its scope and to allocate the river water more equitably, the CFA opens up the possibility for all countries to use the Nile for hydro-development projects. Six of the upstream countries have signed the agreement (as of October 2014), but Egypt and Sudan continue to oppose a few key clauses. During the panel discussion at the World Water Forum, Ethiopian participants brought up their frustration with the downstream countries' intransigence. "For over ten years we have negotiated [with] the CFA," said a representative from Ethiopia's Ministry of Foreign Affairs. "Finally we decided for it to be signed. Six have signed, and we expect the rest to come on board." The Egyptian panelist, who holds a senior position in Egypt's water ministry, grew visibly uncomfortable. When it was his turn to respond to the audience's comments, he said, "As for the CFA, I would prefer to discuss it in a closed room. In public, I prefer to focus on cooperation." To him, the public space of an international water conference was not the place to discuss this politically unsettling limit to the Egyptian government's agency over the Nile.

The topics that came up during this panel discussion were illustrative of some of the other elements that will shape Egypt's water futures. To government officials of downstream countries like Egypt and Sudan, the potential impact of Ethiopia's dam and further hydro-development projects if the CFA is ratified is a more immediate concern than climate change, perhaps because the timeline of impacts is shorter or perhaps because it is a more tangible impact and thus easier to conceptualize than a model output. The construction of an upstream dam with a large storage capacity is a clear manifestation of the transfer of control over future flows from those officials to their Ethiopian counterparts.

These ways in which water is tied to human agency are precisely the things that are pushed to the sidelines in the discussions about Egypt's water future that are framed in terms of climate change. It is not that scientists do not understand the import of political power and control. Anyone who works on the Nile is aware of the conflicts between the basin countries and the political sensitivity of data on past, present, and future river flows. Indeed, many scientists have engaged in efforts to model the potential impacts of proposed dam projects. But active engagement in this politics makes scientists uneasy. An American hydroclimatologist who works on the Nile Basin told me several times during a conversation we had that he tries to stay out of politics. He is still working on modeling East African regional climate, but, he said, "I've gone off trying to look at Nile flows, because it's so political." His choice of what to model, therefore, is a decision not just to avoid engaging in politics but to actively disengage from politics. Scientists like him would prefer to stay in their climate reductionist comfort zone.

For those who are working to understand more about the impacts of climate change on Egypt and other countries of the region, science is one domain, politics is another. The need to strengthen linkages between the two is a recurrent theme at international water fora. "Experts are talking to experts," noted one presenter at the Arab forum, "but the absentees are the decision makers who ultimately make the decisions." A World Bank official agreed. "They [the scientists] are doing research," she said, "they're finding things, but they can't get through to policy makers." This argument is driven by an imagination of a rational bureaucratic planner. If only decision makers understood climate change, the reasoning goes, they would take adaptive measures to minimize its risks. In fact, however, what the discussion at the side event demonstrates is that even if government officials fully understood that climate change could lead to a pronounced decline in Egypt's water supply, and many of them may already do, their primary concern is likely to be what is happening in the upstream basin countries. They have a fundamentally different conception of the balance between natural and human agency in shaping the future of their water supply.

Yet climate change remains the hot topic of the moment, and international organizations and funding agencies continue to insist that governments place this consideration center stage. To encourage planning for climate change adaptation, experts talk about the need for increased awareness. This is about raising awareness not only among government officials but also among water users, especially farmers, who are largely absent from the international water fora in which these discussions are taking place. The assumption is that if water users understand climate change they will alter their behavior in some ways. One presenter at the Arab forum argued, "You need a close dialogue between the users—from a farmer and fisherman—to the top of the government." Another remarked, "Climate change is a new topic for the people. . . . We need to build capacity and equip civil society with the in-depth knowledge regarding climate change."

In one presentation at the Arab forum, an Egyptian scientist presented the results of the UNDP-funded Climate Change Risk Management in Egypt project, which ran from 2008 to 2011. A central component of this project was a public awareness campaign on climate change, since, the scientist explained, "people have to know that there is a problem if they are to change [their behavior]. Like a person who is using water wrongly, he has to know." One of the project outputs was a ten-minute film, which was screened at the forum's banquet dinner. The film started with dramatic images of storms, floods, and droughts. "Egypt is not immune to the climate change phenomenon," said the voiceover, "it affects our limited water." The message for water users was that they need to reduce their consumption. Images of a man washing a car and a woman leaving the kitchen tap running flashed up as indicators of the kind of behavior that would hamper Egypt's ability to deal with climate change. For farmers, who use the majority of Egypt's water, the film made a case for the need for "modern" sprinkler and drip irrigation systems, which use less water than the flood irrigation methods commonly employed by farmers. Accompanied by evocative pictures of wide-eyed smiling children, the film concluded, "Climate change and its negative effects is a fact. It's time to put our hands together and alleviate the suffering of our children."

But how likely is this message to resonate with Egypt's farmers? Is a call to change water use practices based on an argument of climate change risk going to be effective? As I discuss in the next section, farmers do not necessarily see the water that flows through their canals as being closely connected to climate. The issue, therefore, is perhaps not so much that farmers are ignorant of climate change but that they view their water resources in a different way.

CANAL FUTURES

One warm day in March 2008 I visited the end of a canal in the western part of Fayoum province with an engineer from the Egyptian water ministry (fig. 5.2). Nile water flowed out of the canal into small irrigation ditches that led into fields of onions and groves of olive trees, backed, just a short distance away, by bare desert cliffs. We stood watching the water pass through the final stretch of its long journey from a rainstorm over East Africa to the fields. A farmer approached us, irate. "The water is terrible, all my onions have been ruined, all the crops are finished!" he shouted. Another farmer joined him. "I have my irrigation at two in the morning, but no water comes," he said angrily.[8] The engineer responded that he had just checked the water gauge in the canal and found it to be at a good level. "That is only today," the farmer replied. "Here downstream we have a big problem. We have been [to the ministry] a lot to complain and nothing happens. They say yes, yes, but then don't do anything. The upstream farmers are taking our water!" The engineer responded that he had heard that they had employed a guard to ensure that the upstream farmers do not use their pumps outside the time that the ministry officially permits them to do so. "Yes, we did," replied the farmer, "but then they just bought better pumps."

When farmers first moved to this area in the 1960s and started to cultivate the land, the water supply was good. Since then, however, the flow through the canal has declined because of increasing withdrawals. A number of upstream farmers have replaced their offtake pipes in the canal bank with pipes of larger diameter. The size of these offtakes determines how much water leaves the canal and flows into the collecting pools where the pumps operate; the larger the pipes, the more water farmers can access. Now nearly half the land at the end of the canal is left uncultivated for lack of water.

5.2: The end of a canal. Photo: Jessica Barnes.

The next stop on our trip was the weir that lies across the start of this canal. We talked with the gatekeeper, who is responsible for opening and closing the weir that controls how much water enters the canal. He was defensive when the engineer suggested that the water in the canal was low at the end. Then he finally admitted that maybe it had been lower the last couple of days. There had been a holiday so they were away from work and could not open the gates.

In this scene, we see a different kind of knowledge about water futures. This knowledge has nothing to do with climate change or mathematical models. It comes, rather, from farmers' and engineers' daily interactions with the water, as they open weirs, allowing water to flow from one canal to the next, or break down dams of earth, channeling water from a ditch onto their fields. The amount of water flowing down a canal section varies on a number of timescales—from days (today versus the previous days of the holiday) to decades (now versus the past, when water was plentiful). But to those working at this scale and experiencing this variability, the fluctuation in water availability does not reflect changing rainfall patterns in the East

African Highlands. Instead, it is a function of the gatekeeper being away on holiday, upstream farmers expanding their offtakes, the introduction of better pumps, and ministry engineers not prioritizing this area in their water allocation plans. Whereas the previous two visions of Egypt's water future homogenize the outlook to a figure of annual Nile flow into Egypt at the Sudanese border, with the implication that any increase or decrease in water availability will be experienced uniformly across the country, this vision does the opposite. The uncertainty in this vision lies in what will happen to that water after it enters Egypt, where it will flow, and who will get to use it.

Thus farmers see their water as coming from a particular constellation of technical apparatus rather than from a larger hydrological system. They know that the water passing through their canals is Nile water, but they do not articulate their concerns in terms of that broader river basin. They look to the nearby main canal, which they refer to as the big canal (*al-bahr al-kabir*), as their source. They concentrate on the present, not looking to a future of potential water scarcity because the current situation is bad enough. They voice their interests at the temporal and spatial scales at which they experience change and at which they feel they have some capacity to act and influence.

So what does climate change mean to farmers whose water supply does not seem to come from a climatic event, rainfall, but from a techno-political event, the opening of a dam or operation of a pump? Unlike the "local peoples" Crate and Nuttall (2009: 9) write about, for whom climate change "is an immediate, lived reality that they struggle to apprehend, negotiate, and respond to," Egyptian farmers and provincial irrigation engineers are not particularly concerned about climate change. It is not something they are talking about and reacting to on a daily basis. They *are* concerned about changes in the water supply, but they link these not to shifting climatic parameters but to the political and technical decisions that act as intermediaries between the water falling from the sky as rain and them receiving it. Their knowledge about the atmospheric dynamics that produce rainfall and feed a river system may be limited, but they have a close insight into what actually happens to this water as it passes through their province. They understand that water futures at the scale of their irrigation district are tied to human agency, and that that agency is differentially distributed within the population. However much or little water enters the country, they know that the most politically connected and influential among them will always receive the water they need.

This demonstrates, therefore, the necessity of understanding climate change impacts on water resources not only in terms of changing patterns of precipitation and evapotranspiration, but also in terms of the technologies and decision-making processes that mediate the flow of water through the environment. Whether or not climate change shifts the physical production of rainfall, Egyptian farmers will still experience water availability as a technical process, generated not only by engineering but also by politics (Barnes 2014). Just as other studies of climate change awareness have noted the tendency of many people to see climate change as something distant and abstract (Orlove et al., this volume; Smith and Leiserowitz 2012), so too it remains for Egyptian farmers. While scientists push the political issues away from center stage to the sidelines for the sake of creating an appearance of impartial expertise in their analyses of Egypt's water futures, farmers bring political issues to the fore in their interactions with water managers. They identify those whom they see as being responsible for reductions in their water supply and for the future of that supply, and shout loudly to make sure their voices are heard.

SCALE AND AGENCY IN WATER FUTURES

Perceptions of how climate change will impact water resources hence vary depending on the scale at which such futures are evaluated. They are tied closely to how people see the balance of human and natural agency in the production of their resources, and where they locate the source of power for control over those resources. From the perspective of an Egyptian farmer, the effect of climate change on his or her water supply is irrelevant and invisible. Equally, the vagaries of canal openings and pump buying are irrelevant and invisible to the climate modeling community. In the partial description of the former, large-scale, gradual statistical trends are of little consequence in light of the variability introduced by local social and political factors. Correspondingly, these local but immediate experiences are of little consequence to the climate modeler, for whom they constitute part of an unexplained variability that cannot be captured in mathematical representations of atmospheric–land interactions (see also Moore et al., this volume).

Of course, this dichotomy is illusory. Individual experience is nested within phenomena at larger spatial and temporal scales, and these larger-scale phenomena become relevant only through human experiences at smaller temporal and spatial scales. Furthermore, these scales of explanation interact with one another and with diverse interests. Farmers' practices of water use in the fields impact how much water flows between the countries of the river basin and, therefore, how changing precipitation-discharge relations under climate change are manifest. Scientists' research on the impacts of global climate change influence international and national interventions in rural Egypt to improve water management. The key point, however, is to draw attention to the linkage between scale and agency. The scale at which the issue of Egypt's future water availability is framed influences understandings of the human and nonhuman variables that will shape that future.

In highlighting the contrasting perceptions of climatic versus human agency in determining water futures across different scales, my aim is not to make a normative statement. It would be difficult to introduce the question of local water distribution politics into basin-wide hydrological modeling studies, just as it would be challenging to try to bring the probabilistic outputs of global and regional climate and hydrological models to bear on field-level water distribution decisions in a meaningful way. My goal, rather, is to sketch some of the limits to climate change as an explanatory frame. To some, climate change is a key factor in shaping future water supplies; to others, it is incidental. Thus while it is important to recognize the significance of climate change, it is also important to ask for whom it is perceived as significant and for whom it is not, and to acknowledge the political ramifications of these distinct stances. Given the increasing international attention being paid to climate change, we expect climate change to be a vital issue everywhere. This is part of the climate reductionism Hulme (2011) identifies as being so prevalent. But in the case of Egypt's water, climate change may not actually be the central factor driving the future of this critical resource. The fact that international attention and funding are focused on climate change impacts on Egypt's water supply is perhaps linked less to the dominance of this issue over others that affect Nile flows and more to the political imperatives of maintaining funding to a geopolitically crucial ally, and to the fact that climate change has become such a buzzword within the donor community. Climate change in Egypt, so far as many international experts are concerned, is about water, but water in Egypt is not just about climate change.

NOTES

I would like to thank participants of the workshop on Climate Change and Anthropology held at Yale in April 2012 for their feedback on an earlier draft of this paper, in particular the detailed comments of Frances Moore and Michael Dove.

1. I distinguish between five sets of actors for the sake of clarity, although these categories overlap. The first group, whom I refer to as *scientists*, comprises people who work in the modeling departments of national water ministries, research institutes, and universities. Many of these scientists have doctoral degrees, often earned in Europe or the United States, sometimes in climatology, other times in engineering or related disciplines. The second group is *government officials*, who hold more senior positions in the water ministries and who have power over decision making. Often these people are also scientists and hold doctorates, but they are distinct from the first group in that they are more directly engaged in the political realities of policy making and answering to the elected authorities. The third group I term the *international experts*, who are non-Arabs working on issues related to climate change in the Middle East. I use the term *expert* rather than *scientist* not because I see these people as having more expertise than the other groups, but in part because they often see themselves in those terms and in part because the group includes both those who are focused on policy (development practitioners) and those who are more scientifically focused (hydrologists, climate specialists, engineers). The fourth group, the *irrigation engineers*, are the people who work for Egypt's Ministry of Water Resources and Irrigation at a provincial level and are involved with the everyday management of the water distribution system. Finally, I refer to *farmers* as a group, while acknowledging that there is great variability within this category.
2. While I use the term *scale* without qualification, I do so with the understanding that scale is not a given but rather is produced in and through particular relations of power (Smith 1992).
3. The president of the Arab Water Council is a former minister of water in Egypt, and several other members of the council's executive committee hold or formerly held senior positions in the ministry.
4. While this scientist talked in dichotomous terms of scientists versus policy makers, as I note earlier in the chapter these groups are not distinct, nor are their perspectives. There are some contexts in which policy makers engage scientists in discussions about climate change, but others in which, as I describe in the next section, their priorities lie elsewhere.
5. The assertion that the dam's storage capacity will benefit downstream riparian countries is based on the fact that evapotranspiration rates in the Ethiopian Highlands are significantly lower than those in Sudan and Egypt. Hence if water from the seasonal rains over the highlands is stored in this part of the basin and gradually released to sustain year-round flow through the Blue Nile, less will be lost to the atmosphere and more will flow on down into the Nile.
6. *Green water* is a term used by water managers and scholars to refer to rainwater that infiltrates and is stored in the soil, in contrast to the water stored in surface rivers and lakes and groundwater, which they call *blue water* (Falkenmark and Rockström 2006).
7. For a detailed account of this history of contestation over Nile waters, see Waterbury (1979, 2002) and Howell and Allan (1994).
8. The ministry allocates water to farmers in this region according to a rotation system, whereby farmers are entitled to irrigate their land for a set period of time each week.

REFERENCES CITED

Ahearn, L. 2001. "Language and Agency." *Annual Review of Anthropology* 30: 109–38.

Barnes, J. 2014. *Cultivating the Nile: The Everyday Politics of Water in Egypt.* Durham: Duke University Press.

———. forthcoming. "Uncertainty in the Signal: Modeling Egypt's Water Futures." *Journal of the Royal Anthropological Institute.*

Crate, S., and M. Nuttall. 2009. *Anthropology and Climate Change: From Encounters to Action.* Walnut Creek, Calif.: Left Coast Press.

Dove, M., A. Mathews, K. Maxwell, J. Padwe, and A. Rademacher. 2006. "Questions of Agency in Current Conservation and Development Discourse." In *Against the Tides: The Vayda Tradition in Anthropology and Human Ecology*, ed. B. McCay and P. West, 225–54. Walnut Creek, Calif.: Altamira Press.

Falkenmark, M., and J. Rockström. 2006. "The New Green and Blue Water Paradigm: Breaking New Ground for Water Resources Planning and Management." *Journal of Water Resources Planning and Management* May/June: 129–32.

Gesellschaft fur Internationale Zusammenarbeit (GIZ). 2011. *Water and Climate Change in the MENA-Region: Adaptation, Mitigation, and Best Practices:* Deutsche Gesellschaft fur Internationale Zusammenarbeit. Documentation of conference on Water and Climate Change in the MENA Region, Berlin, Germany, April 28–29.

Howell, P., and T. Allan. 1994. *The Nile: Sharing a Scarce Resource: An Historical and Technical Review of Water Management and of Economical and Legal Issues.* Cambridge: Cambridge University Press.

Hulme, M. 2011. "Reducing the Future to Climate: A Story of Climate Determinism and Reductionism." *Osiris* 26: 245–66.

Jasanoff, S. 2010. "A New Climate for Society." *Theory, Culture and Society* 27: 233–53.

Jonsdottir, A. 2013. "Scaling Climate: The Politics of Anticipation." In *The Social Life of Climate Models: Anticipating Nature*, ed. K. Hastrup and M. Skrydstrup, 128–43. New York: Routledge.

Smith, N. 1992. "Contours of a Spatialized Politics: Homeless Vehicles and the Production of Geographical Scale." *Social Text* 33: 54–81.

Smith, N., and A. Leiserowitz. 2012. "The Rise of Global Warming Skepticism: Exploring Affective Image Associations in the United States Over Time." *Risk Analysis* 32: 1021–32.

Sowers, J., A. Vengosh, and E. Weinthal. 2011. "Climate Change, Water Resources, and the Politics of Adaptation in the Middle East and North Africa." *Climatic Change* 104: 599–627.

Tolba, M., and N. Saab. 2009. "Arab Environment Climate Change: Impact of Climate Change on Arab Countries." Report of the Arab Forum for Environment and Development, Beirut, Lebanon.

Tsing, A. 2005. *Friction: An Ethnography of Global Connection.* Princeton: Princeton University Press.

Vidal, J. 2011. "Drier, Hotter: Can Egypt Escape Its Climate Future?" *Guardian, Enviroment Blog.* November 18. http://www.theguardian.com/environment/blog/.

Waterbury, J. 1979. *Hydropolitics of the Nile Valley.* Syracuse: Syracuse University Press.

———. 2002. *The Nile Basin: National Determinants of Collective Action.* New Haven: Yale University Press.

Chapter 6 Satellite Imagery and Community Perceptions of Climate Change Impacts and Landscape Change

Karina Yager

Climate change is a scientific and conceptual term that has gained global recognition and reflects an increasingly shared understanding of the rapid change of Earth's climate systems. However, the causes, definitions, and outcomes of climate change are not universally agreed upon. Concepts of climate change draw upon a diverse range of data and observations that may be considered spatially and temporally relative and culturally or geographically specific. Thus, understandings of climate change in one community or society may be quite different from those in another, in part depending on shared knowledge sets and experience. Furthermore, insights into the causality and consequences of local and global environmental change are complicated by the fact that climate change is being accompanied by rapid societal changes, including population growth, migration, urban sprawl, and other human dimensions of natural resource use.

In the Andes of South America the impacts of recent climate change are relatively well documented by scientific investigations, including studies of glacier melt (Francou et al. 2000; Georges 2004,

Rabatel et al. 2013; Soruco et al. 2009), temperature increase (Diaz and Graham 1996; Mark and Seltzer 2003; Urrutia and Vuille 2009; Vuille et al. 2003; Vuille and Bradley 2000), increase in tropical freezing level heights (Bradley et al. 2009; Diaz et al. 2003), and species migration (Seimon et al. 2007). These changes directly affect Andean societies, resulting in environmental hazards (Carey 2008), water resource scarcity (Bradley et al. 2006; Coudrain et al. 2005; Mark and Seltzer 2003; Vuille 2007), critical loss of hydroelectrical power and agricultural production (Vergara et al. 2007), and impacts on biodiversity and traditional production practices (Young and Lipton 2006).

In addition to scientific investigations of climate change, research with local communities in the Andean region has revealed the value of these communities' knowledge and experience in identifying the trends and impacts of climate change (Orlove et al. 2000; Rhoades et al. 2006). For example, participatory workshops with community members have resulted in the identification of real-time climate data that are often overlooked or unidentifiable by scientific instrumentation or measurement alone (Ulloa and Yager 2007). With increasing attention turning to the value of "local ecological knowledge" or "traditional ecological knowledge" within the policy and development community (Becker and Ghimire 2003; Berkes 1999; Chalmers and Fabricius 2007; Gadgil et al. 2003; Puri, this volume), important questions arise, such as, How do local knowledge datasets on climate change compare to scientific datasets? What are some of the ways in which the former complements the latter, and where are the points of contradiction?

This chapter explores these questions through a comparison of current scientific research and local knowledge on climate change impacts in Sajama National Park, Bolivia. I focus on the impacts of climate change on peatlands, which constitute a critical resource for herding communities living in this area. I compare a satellite image analysis of peatlands within the park with local observations of landcover change. I also consider the difficulty of distinguishing climatic from social drivers of change in peatlands. In doing so, I demonstrate how scientific data and local ecological knowledge offer both complementary and sometimes divergent assessments of landcover change. Both are valid in their own way, but each has limitations. Considered together, they provide a more holistic understanding of the dynamics and significance of landcover change at the local and regional level.

ANDEAN PEATLANDS AND THEIR SOCIO-ECOLOGICAL IMPORTANCE

Peatlands are recognized as being vital to both natural systems and human populations, playing a crucial role in climate, biodiversity, and society (Strack 2008). They constitute an estimated 3 percent of Earth's total land-cover; the vast majority are found in the boreal and subarctic regions of the Northern Hemisphere, covering expansive areas (circa 350 million hectares [ha]) in North America, Russia, and Europe. The remaining 10 to 12 percent of global peatland cover is in the tropics, including expansive lowland peat systems in Southeast Asia and the Americas. In the Andes of South America, a unique type of alpine peatland occurs, locally identified as *bofedales*, which provides key environmental services that support Andean mountain biodiversity and high-altitude human populations, especially pastoral societies.

Alpine peatland systems occur across the range of Andean ecosystems, from the *páramo* (humid grasslands) of Ecuador and northern Peru (Bosman et al. 1993; Chimner and Karberg 2008), to the *jalca* (transition zone) of central Peru (Cooper et al. 2010), to the drier *puna* (semiarid to arid grasslands) region of southern Peru, Bolivia, and northern Chile (Campos and Yancas 2009; Earle et al. 2003; Ruthsatz 1993, 2000; Squeo et al. 2006). In the semiarid to arid tropical Central Andes, alpine peatlands often appear as green oases, composed of tightly compact cushion plants, interwoven with hydrological systems and often situated downstream from glacier peaks. Found at high elevations, typically from 3,500 to 5,200m, alpine peatlands occur in a variety of geomorphological settings, including glacier valley basins and steep lateral mountain slopes. The size of individual systems varies greatly, ranging from a few hectares to several thousand, depending on landscape factors such as geology, hydrology, and anthropogenic influences (Squeo et al. 2006). Peatlands are also referred to, depending on their primary hydrological characteristics, as cushion mires, peat-accumulating fens, or highland bogs and vary ecologically in terms of species richness, organic content, and chemistry (Cooper et al. 2010).

Of particular relevance to today's global climate systems, peatlands around the world play an important role in the global carbon cycle (Dise 2009; Gorham 1991; Tolenen and Turunen 1996). Peatlands sequester carbon and are a known source of greenhouse gasses, including methane (Moore 1989). Environmental perturbations, especially those related to fluctuations in the water table, directly affect carbon accumulation and greenhouse gas emissions from peatland systems (Dise 2009; Earle et al. 2003; Roulet et al. 1992).

Although Andean peatlands constitute a smaller total area compared to northern latitude peatlands, some studies indicate that the former have significantly higher rates of carbon accumulation than the latter per unit area (Chimner and Karberg 2008; Earle et al. 2003). Because alpine peatland systems are located in a tropical environment, peat accumulation occurs on a year-round basis, which contributes to rapid carbon cycling and significant carbon storage on a local and regional scale. For example, preliminary studies indicate that one square meter of peatland (dominated by *Oxychloe*) sequesters the same amount of atmospheric carbon in one year as at least ten square meters of *Sphagnum* peatlands (typical of the Northern Hemisphere) (Earle et al. 2003: 10). It follows that if Andean peatlands covered the same amount of total landcover as their northern counterparts, there would be notably higher amounts of carbon sequestration on a global scale. Peatland systems in the tropics are often several meters in depth, storing over a millennia of peat accumulation, and their role in carbon sequestration is locally and regionally significant, especially compared to other landcover types.

Andean peatlands support tropical mountain biodiversity. They can act as a natural water storage support system, through the capture of seasonal precipitation and glacier meltwater outflow. Given the high concentration of biomass and water resources, peatlands are hotspots of mountain biodiversity, hosting many endangered and endemic species. Many highland species rely on alpine peatlands for food and water, including the wild, endangered vicuña (*Vicugna vicugna*) and numerous Andean birds.

Andean bofedales are especially critical for herding communities, who rely on them to support pastoral production. They are a key resource for extensive wetland-dependent camelid production, providing dense humid vegetation that serves as high-quality forage in an otherwise seasonally arid and sparsely vegetated mountain environment. At high elevations in the Andes, the extreme mountain climate is characterized by freezing temperatures on a year-round basis, which limit the ability of communities to practice agricultural production, with the exception of small-scale tuber production. Given the mountain setting, pastoralism is an alternative, ecologically viable means of production (Browman 1987). Andean pastoralism is traditionally based on the herding of the llama (*Lama glama*) and alpaca (*Vicugna pacos*) and is an integral livelihood practice of Andean society that has spanned several millennia to the present (Baied and Wheeler, 1993; Browman 1989; Flores-Ochoa 1977; Murra 1965; Orlove 1977). In the tropical Central Andes, current estimates of livestock include a combined population of over 2.8 million camelids in Bolivia and

5 million in Peru (UNEPCA 1999), many of which exclusively graze at high altitudes and therefore rely on peatlands as a primary source of forage. Today, many pastoral communities living in the puna manage peatlands through irrigation to maintain and expand the area of pastures for their animals (Alzérreca 2001; Buttolph 1998; Palacios-Ríos 1992; Yager 2009). Thus, peatlands are anthropogenic components of Andean mountain ecosystems and are integral to Andean pastoral production.

SAJAMA NATIONAL PARK

Established as the first national park in Bolivia in 1939, Sajama National Park is located in the Cordillera Occidental of the Andes, in an arid to semiarid climate characterized as the dry puna region (Troll 1968) (fig. 6.1). The atmospheric circulation regime produces distinct wet and dry seasons in the region, with 80 to 90 percent of the total annual precipitation occurring during the austral summer, that is, the short, wet period between December and March (Hardy et al. 1998; Vuille 1999; Vuille et al. 1998). Unfortunately, long-term meteorological measurements for the park are unavailable owing

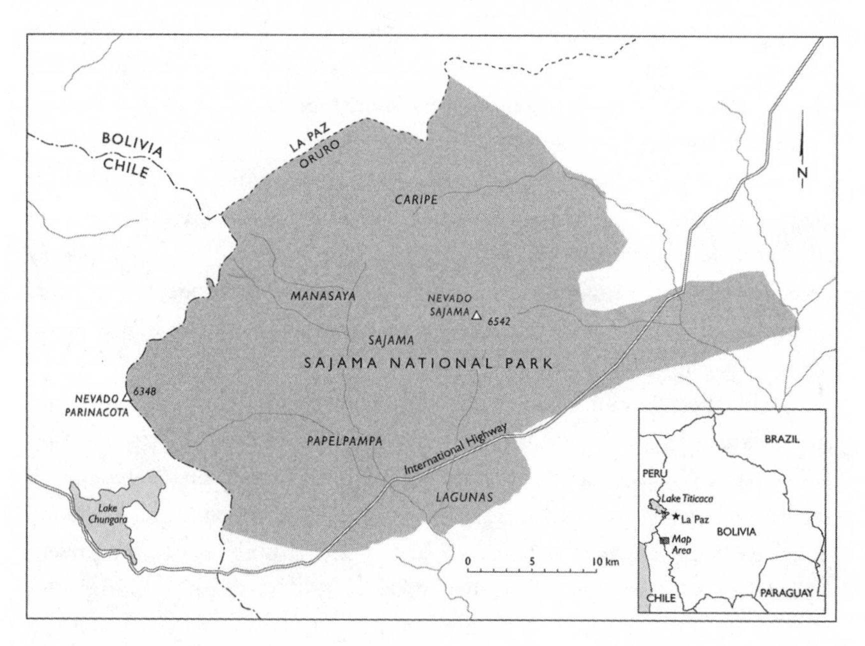

6.1: Sajama National Park. Bill Nelson.

to the lack of consistent local instrumentation, a challenge for long-term climate studies that exists across the high Andes (Diaz and Graham 1996) and also limits local scientific monitoring of recent climate change trends related to precipitation and temperature. According to one local, short-term study conducted at the base of Nevado Sajama (approximately 4,200m), the average annual temperature is 3.4° C with a mean annual precipitation of 347mm (Liberman-Cruz 1986). A weather station located 110 km north-northwest of the study area, in Charaña, at 4,057m altitude, records an average annual temperature of 4.7° C and annual average precipitation of 300mm over a twenty-year period, from 1986 to 2006 (Servicio Nacional de Meteorología e Hidrología, La Paz, Bolivia).

The park has the highest altitude forest in the world, composed of the Andean *queñua* (*Polylepis tarapacana*), with a treeline up to 5,200m. Nevado Sajama (6,542m) is the geographic centerpiece of the park and contains the oldest glacier climate record in the Andes, recording 25,000 years of climate history in its ice cap (Hardy et al. 2003; Thompson et al. 1998) (fig. 6.2). Nevado Sajama is also an important mountain deity (*apu*) for indigenous highland communities.

6.2: Nevado Sajama. Photo: Karina Yager.

Management of the park is a collaborative decision-making and planning process among community members, national park administration, and local municipalities who meet on a monthly basis (Hoffmann 2007). There are five indigenous Aymaran communities (Caripe, Lagunas, Manasaya, Sajama, and Papelpampa) residing in the park that identify with the pre-Hispanic and pre-Inca Karangas culture. The majority of local residents practice pastoral production, but many also work in tourism and transportation. The number of full-time permanent residents fluctuates, as many emigrate seasonally for education and employment opportunities. Local census estimates range from 764 to 1,700 persons on a full-time basis, the majority of whom are children, youth, and elders (Espinoza 2001; MAPZA 2001). An international highway constructed in 1994 intersects the southern perimeter of the park and connects the urban centers of La Paz, Bolivia, with Arica, Chile, thereby facilitating the transinternational movement of people and products through the *altiplano*. Agriculture is not a viable means of local production, in part because of the arid climate and cold temperatures. A few families grow hardy tubers for subsistence in sheltered areas or unused corrals, but the majority purchase agricultural goods from regional markets or bring in goods from La Paz.

Within the protected area, totaling around 1,000 km^2, the native pastures provide sustenance for approximately 50,000 domesticated animals. Llama make up the majority of herd composition, totaling 49 percent (24,170), followed by the alpaca with 43 percent (21,318), and then sheep with 8 percent (3,828) (MAPZA 2001). Herd composition reflects regional economic production, wherein the primary product sold at local markets and to buyers is llama meat, followed by alpaca wool (Acarapi-Cruz 2007). *Charqui*, dried camelid meat, is also locally produced. Sheep are predominantly raised for local consumption and rapid sale. Donkeys and horses in very small numbers are also raised in the park, mostly for trekking services. In addition to livestock herds, a principal grazing animal in the park is the vicuña. The vicuña has made a rapid recovery in recent decades owing to community efforts to protect it and sustainably harvest its wool. The estimated population is around 3,300 animals in the park (Maydana 2010).

PEATLANDS (BOFEDALES) IN SAJAMA NATIONAL PARK

Peatlands, locally termed bofedales in Spanish and *oqho* in Aymaran, are one of six primary vegetation classes within the park: semiarid, high-Andean vegetation (*semi-desierto altoandino*), forest (*queñuales*), shrubs (*tholares*), dry grasslands (*pajonales*), peatlands (bofedales), and salt flats (*colpares*) (Beck et al. 2010; Yager 2009). The bofedales and pajonales are considered important production zones (Mayer 2001), actively managed by local communities. The bofedales are the most important vegetation class and production zone for pastoralism because they have the highest biomass of any other vegetation class in the puna, providing a source of highly palatable, nutritious forage for livestock on a year-round basis. Pastoralism, as it has existed over several generations to millennia, in this high-altitude mountain environment would not be sustainable without healthy peatland systems.

Bofedales in the park are primarily dominated by vascular plants from the Juncaceae family and characterized by mixed vegetation associations that include a variety of species typical of wet meadows, including the families of *Asteraceae, Cyperaceae,* and *Gramineae* (Beck et al. 2010). The estimated consumable biomass ranges between 880 and 9,042 kg dw (dry weight) per hectare (Alzérreca 2001). This is considerably more than the secondary grazing zone–the dry grasslands (*pajonales*), that are dominated by tussock grass *(Festuca orthophylla)* and other inter-tussock vascular plants, which have a mean total biomass of 1.08 ± 0.7 g dw per individual plant species (including leaves, stem, storage, and roots) (Patty et al. 2010). Inter-tussock plants, though nutritious and palatable, have a very limited total landcover, which correlates with annual precipitation patterns and grazing intensity (Yager et al. 2008a).

Bofedales may be broadly classified according to a gradient of humidity (Alzérreca 2001; Buttolph 1998), with humid, or primary, bofedales typically dominated by *Oxychloe andina* and *Distichia muscoides,* whereas the more xeric, or secondary, bofedales may be dominated by *Plantago tubulosa* and *Carex sp.* and associated with grasses like *Deyeuxia rigescens* and *Deyeuxia spicigera.* The primary bofedales remain mostly humid year-round, while secondary bofedales experience more excessive seasonal drying (Alzérreca 2001; Buttolph 1998). The water pools of bofedales typically contain such species as *Potamogeton strictus, Myriophyllum quitense, Nostoc sp.*, and *Ranunculus sp.* According to local herders, the humid bofedales are the favored grazing zone of alpacas, while llamas tend to prefer grazing in the dry grassland zones. The Aymaran name for the two main bofedal plants is *paco;* the *Distichia* is considered a female plant (*kachu paco*) and the *Oxychloe* male (*orko paco*). The latter also produces a small fruit that is edible. Dry grasses, such as *Poaceae (Gramineae*), which grow in and near bofedales are locally called *oqhosiki.*

The pastoral counterpart to the bofedales are salt flats, locally termed *colpares*. Colpares in the park consist of species that are highly tolerant to salt concentrations, including *Sarcocornia pulvinata, Frankenia triandra*, and *Atriplex nitrophiloides*. Local herders recognize that the salt content of colpares provides an important nutritional requirement for livestock animals. Despite the nutritional value of colpares, the total plant cover in colpares is very low, typically less than 20 percent. Livestock graze here occasionally, but colpares meet only a supplemental need and do not qualify as a significant pasture zone.

LOCAL OBSERVATIONS OF CHANGE

In 2007 a participatory workshop was held to share, discuss, and identify the current observations and understanding of climate change at the local level. The workshop took place in Sajama National Park and brought together seventy-five individuals, including local community members, park personnel, park researchers, students, NGO representatives, scientists, and policy makers from local and national agencies. The aims of the workshop were to: (1) identify and discuss the perspectives of climate change experienced by local community members in the park; (2) present and discuss scientific research related to climate change, including local and regional studies, with local communities; (3) identify possible strategies of adaptation to climate change and to other social and environmental changes; and (4) raise the awareness of youth in the park on the subject of environmental change (Ulloa and Yager 2007; Yager et al. 2008b).

In this workshop the sustainability of pastures in the face of climate change was identified as a primary concern by the local community members. Glacier loss was also a major worry, as the retreat of Nevado Sajama has been very rapid (Yager 2009). Through various workshop exercises, including focus groups, timelines of landscape change and climate events, and community mapping and modeling, a consensus was reached on bofedales, one which emphasized the loss and degradation of healthy pastures for animals in recent decades (Ulloa and Yager 2007; Yager et al. 2008b). Many said the bofedales were drying up and stated that vegetation health and plant biodiversity (that is, species richness) were declining. One group of participants summarized the general view on bofedales by stating, "There used to be a lot of grass and pastures because it rained at the right time. Now they [the grasses] don't grow anymore; some areas have temporarily dried up and there is insufficient food for livestock. There is a larger expanse of the colpares [salt flats] than there used to be."

The degree of degradation was further demonstrated in community mapping of the pastures (fig. 6.3). Participants mapped the relative condition of the pastures, using three colors to symbolize the various qualities of pasture. Green represented healthy pastures that typically stayed green year-round, red represented pastures that were more susceptible to seasonal drying or that were in the processes of drying completely, and black represented areas that were excessively dried or had become permanent colpares (fig. 6.4).

The participants described aspects of vegetation change, with reference to plants, pastures, and ecological zones of production. One group said that particular bofedal plants, namely, *paco* (*Oxychloe andina*) and *pork'e* (*Deyeuxia sp.*), no longer grew in certain areas. They also described primary pasture areas as transforming to dry pastures, or *puro pajonal*, where dry tussock grass (*Festuca orthophylla*) dominates. Many mentioned that the colpares have expanded in recent decades, perhaps because of a lack of water. Many of the bofedales in the maps were described as once having been greener and more expansive but now being degraded and in danger of drying owing to the drying of rivers and springs. Many talked about the multiple

6.3: Community mapping workshop. Photo: Karina Yager.

6.4: Community mapping workshop. Photo: Karina Yager

indications that the climate was changing, including rapid glacial loss, less rain during the wet season, longer frosts during the dry season, greater occurrence of prolonged droughts, and unexpected extreme events, both high-heat days and unseasonal freezes. The prolonged drought periods were of great concern in relation to pastures. Bofedales are water dependent, and excessive drying means that these systems will die. Many participants also believe that when the bofedales dry out completely they become colpares.

The inhabitants of Sajama National Park depend on the peatland systems to support local livestock production, and they reported that it is increasingly challenging to keep these vital pastures green. To many, this problem was linked to climate change. After the workshop, a case study was developed to evaluate peatland change in the park using satellite image analysis. Specifically, the study compares change in the extent of bofedales and colpares, comparing the total landcover of each over a twenty-year period, between 1986 and 2007.

SATELLITE ANALYSIS OF CHANGE

The satellite image analysis included use of Landsat data (Path 01, Row 73) with 30-m resolution, acquired from the GLOVIS facility at USGS-EROS. This study of interannual change compares bofedales and colpares during 1986 (July 30) and 2007 (July 24), each at the peak period of the dry season. Forage is most scarce at this time of year (Patty et al. 2010; Yager et al. 2008a). It is also a time of year when the livestock are highly susceptible to illness, malnutrition, and even death if not sufficiently nourished (Alzérreca et al. 2006).

Images were geo-referenced to the same datum (WGS84) and projection (UTM Zone 19S) and subset to the area of study, Sajama National Park. Image preprocessing steps included atmospheric correction and the conversion of digital numbers to at-satellite reflectance to calibrate for differences in illumination and path radiance during image acquisition (Chander and Markham 2003). A digital elevation model (DEM) was created with combined datasets from Shuttle Radar Topography Mission (SRTM), acquired from Global Landcover Facility at the University of Maryland, and GTOPO30 project from USGS-EROS. The SRTM product (WRS-2 filled Finished B) with 90-meter grid posting (3 arc second) serves as the primary base data for the DEM, while data gaps occurring at high elevations were filled with reference to the GTOPO30 dataset. The bofedales and colpares are primarily concentrated in the altiplano region at the base of Nevado Sajama, around 4,200m in elevation, with some extending up to 4,700m. A layerstack of data was created for each acquisition date with four bands of information: band 3 (0.63–0.69 μm, corresponding to visible red reflectance), band 4 (0.76–0.90 μm, corresponding to near-infrared reflectance), band 6 (10.4–12.5 μm, corresponding to thermal infrared), and the DEM.

To evaluate landcover change between the images, I applied the Normalized Difference Vegetation Index (NDVI), a ratio based on the reflectance of near infrared radiation and red radiation, which is commonly used for monitoring and evaluating vegetation change (Tucker 1986). The bofedales are compact vegetation systems with a reflectance in the near infrared that significantly exceeds all surrounding vegetation classes, such as the *queñua* forests or dry grasslands, which have a sparser total vegetation cover and thus a significantly lower NDVI than peatlands. The mean NDVI value of bofedales has a range between .35 and .37 in the area of study, and a threshold value was selected to isolate the landcover class of bofedales, and the total area of peatlands was then calculated. The study of change focused on the primary bofedales—the humid vegetation of peatland pastures that is most critical for sustaining pastoralism year-round.

Colpares, in contrast, are primarily composed of salt minerals and therefore have a very high reflectance, corresponding to high brightness values in the visible range. The only other landcover class with comparable reflectance is snow cover and glacier, which occur at elevations above 5,000m in the park. Since the bofedales and colpares within the park are located in the flat altiplano basin that is located well below the glacierized areas and steeper lateral slopes, the DEM, combined with a brightness threshold, could be used to isolate the colpares in the image. Thus a mask for brightness values and elevation was applied to isolate and then calculate the total area of colpares.

Through this methodology, the satellite image analysis revealed an increase of 2.48 percent in the total number of pixels with high NDVI, indicating an increase of 117 ha in the total area of primary bofedales between 1986 (4,779 ha) and 2007 (4,896 ha). A greater change was detected in the total area of colpares, which increased significantly in size, by 36.8 percent, or 792 ha between 1986 (2,151 ha) and 2007 (2,943 ha). Based on image inspection, the lower-grade pastures or mixed vegetation zones surrounding peatlands, which are more susceptible to drying, are severely eroding or converting to salt flats, while the total area of primary bofedales has increased by a small percentage

SOCIO-ECOLOGICAL DRIVERS OF LANDCOVER CHANGE

Understanding landcover change in Andean peatland systems requires a consideration of multiple datasets, including interdisciplinary research on the socio-ecological factors that influence production dynamics (Buttolph and Coppock 2004; Coppock and Valdivia 2001; Rocha and Sáez 2003), research on climate change processes in the Andes (Vuille et al. 2003), and an examination of the relationship between peatlands and mountain geohydrology at individual sites (Campos and Yancas 2009; Earle et al. 2003; Squeo et al. 2006). Contributing to this research, satellite image analysis is a valuable means of evaluating and monitoring the dynamics of socio-ecosystem components in the Andes, such as the peatlands presented herein as well as glaciers (Kaser 1999; Silverio and Jaquet 2005) and Andean vegetation (Postigo et al. 2008; Washington-Allen et al. 1998). In addition, local ecological knowledge helps bring to attention concerns about landcover change and to identify both social and climate-related trends and drivers of change.

Peatlands are highly sensitive to change in hydrological inputs, being dependent on local water quantity and quality, including factors such as chemistry, level, flow, temperature, and seasonality. Peatland systems that are hydrologically linked to glacier outflow will be directly affected by recent, rapid deglaciation processes. A temporary increase in glacier outflow, especially during the dry season, may help to sustain or increase water levels available to irrigate (naturally or through management) peatland systems. Similar processes have been studied in other regions of the Andes, where image analysis also found an increase in wetland areas in recent years (Postigo et al. 2008). Other studies indicate that long-term climate fluctuations influence the rate of accumulation and upstream migration of individual Andean peatland systems (Earle et al. 2003). Across the Andes rapid deglaciation trends may result in a short-term expansion of peatland systems in some localized places but result in desiccation of peatland systems over the long term by decreasing total water availability for irrigation. Further research is needed to identify the key hydrological systems that sustain peatlands and how these sources will be affected by climate change.

The continued maintenance of total peatland area in the park is influenced by both human management practices and recent environmental changes. This localized remote sensing study in Sajama National Park finds a small degree of change in the total area of peatlands within the park from 1986 to 2007, but a more significant change in the extension of colpares, or salt flats, over the same period. Local ecological knowledge voiced by community members had held that the peatlands were drying, but in fact the healthy peatland communities actually increased in total area over the twenty-year study. This does not mean that local knowledge is incorrect, as the image analysis compared only total cover, not specific pastures. The small increase may be due to normal variance in NDVI on an interannual basis. Precipitation records for each year did not indicate a significant difference.

However, locals had expressed that some pasture areas were drying, and this was indeed shown to be true by the comparison of landcover change of colpares. Results from the image analysis find the total area of the greenest, or primary, bofedales has been relatively maintained but that seasonally drying meadows and xeric, mixed-pasture communities, above all, salt flats, increased. Therefore, local knowledge was not erroneous but instead encompassed broader conceptions of total pasturelands or referred to particular local pasture sites. Indeed, the assertions of overall forage desiccation were confirmed by the sizable increase in colpares, as detected in satellite image analysis. Future remote sensing analysis would be useful to identify the location of pastures that remained healthy and to seek to determine why. Visits with local community members to peatland sites represented in the satellite imagery and community maps and a more in-depth investigation of both the ecological and social changes impacting pasture health would also be valuable. A remote sensing study linking peatland species communities with their spectral signature as evaluated on an interannual basis would help to characterize the level of degradation or shift in landcover occurring in recent years at specific pasture sites. Further research is also needed on identifying pasture specific drivers of landcover change. Additional research to address the knowledge gaps on local climate change processes (Mohr et al. 2014) and peatland change (Slayback et al. 2012; Cooper et al., submitted) is currently under way.

Local causes of salinization primarily result from a change in the amount of available water and excessive accumulation of salts in soil (Cardozo et al. 2004), which may result from shifts in groundwater amounts, changes in volume due to precipitation or seasonality of water flows, among other possibilities related to climate change impacts. Salt concentration in the root zone of hydromorphic plants, such as in the bofedales, restricts the ability of the vegetation to absorb water, and, if excessive, leads to severe dehydration of the plant. Salt deposits on the soil act as a cap that restricts further drainage of water into the soils. Depending on local topography, horizontal surface flow results in the distribution of salt concentrates across the surface and, with evaporation, further extends the salinization process. Other contributing factors may include anthropogenic activities, such as the diversion of water for irrigation to maintain hydrological flows in nearby higher-quality peatland pastures and overgrazing of pastures.

In addition to the possible impacts of climate change, herd management practices directly influence peatland systems, affecting total vegetation cover, species richness, and soils. These practices include decisions about herd composition, pasture rotation, and livestock numbers, all of which directly affect pasture quality and the sustainability of grazing. In Sajama National Park the llama and alpaca are relatively low-impact grazers compared to introduced species, particularly sheep (Baied and Wheeler 1993; San Martin and Bryant 1987). Camelids are better suited for the mountain ecosystems because they have padded hooves, a prehensile upper lip for forage clipping, and efficient digestion of puna grasses. Sheep, in contrast, typically tear into soils and rip plant material from the roots when grazing. Though sheep compose less than 10 percent of total animals in the park, their impact is nonetheless considerable, and efforts are made to keep the number of sheep in the park to a minimum. The greater concern in Sajama National Park is not herd composition per se, given that the majority are native camelids, but the high number of animals grazing in any given area for extended periods of time. Pasture rotation is becoming rare in the face of increased fencing in the park, which is used not only to privatize pasture zones but also to exclude neighbors' animals and decrease the attention required to monitor animals' movements. Coupled with a steady increase in herd animal numbers per each family member, the average carrying capacity in the park has already been largely exceeded.

Recent social change is an additional factor affecting pasture quality and resilience. Population increase, out-migration, change in modes of transportation and exchange of goods, all contribute to significant change in the pastoral management practices in the park. The lack of access to land for new generations results in an increase in total herd numbers grazing on individual *estancias,* or ranching zones. The local population comprises primarily elders and children, the majority of those in the middle-age range having left for education and job opportunities. Out-migration has resulted in elders being left to care for herds and in the hiring of part-time herders. Hired help typically does not invest in pasture maintenance, and elders are often not physically able to manage herds as effectively. Given the new perceived opportunities available in urban centers, few pastoral households have enough capable persons to devote to herd management and pasture improvement. Historically, pastoralists traveled with their herds to exchange goods, which allowed an average of three months of pasture fallow. Today, goods are purchased at market, and camelid caravan travel ceased around the middle of the twentieth century. The outcome of these recent social changes includes higher herd numbers and deterioration of pasture rotation, which directly influences the health and sustainability of pastures.

As a final exercise in the participatory workshop, there was a group discussion of possible adaptive actions to help deal with the identified landscape and climate change trends. Considering the local issues of water, livestock, and bofedales, the participants identified several options for both community and individual action. Regarding peatlands, the suggestions primarily centered around increasing and maintaining the hydrological levels of bofedales, including rescheduling the timing of irrigation to commence earlier in the year, elevating riverbeds that have been eroded to achieve greater water availability, rotating irrigation beneficiaries more evenly on a community basis, and creating wells to increase water supply for irrigation. Other suggestions included a need to emphasize the quality of herd animals over quantity, to develop greater community zoning of pasturelands to avoid competition and overgrazing, and to increase reliance on vicuña management over traditional pastoral management.

FUTURE OF PEATLANDS AND PASTORALISM

Change in Andean peatland systems is the product of multiple human and natural processes over time. Various socio-environmental drivers of bofedal change may be considered, including shifts in climate, herd management, pasture management, change in hydrological sources of peatlands, and recent social change. In Sajama National Park community members recognize that climate change may threaten the sustainability of bofedales in the long term, which would have direct impacts on local livelihoods that depend on pastoral production, but they also recognize the role that recent social trends and local management have on peatland systems.

The case study presented here shows that Andean peatlands are dynamic ecosystems that respond to changes in the local environment caused by a combination of anthropogenic activities and climate change processes. Over the past twenty years the small increase in primary peatlands may be partially due to natural interannual variation in vegetation cover, or a combination of local irrigation efforts and possible short-term increase in the availability of water (with recent deglaciation or hydrological shifts) at very localized sites. Although locals believed that the majority of peatlands were declining, according to the image analysis, they are being maintained at a regional park level. The question is perhaps not total area per se, but pasture condition, which needs further study in the field. While the overall area of primary peatlands may be sustained, many marginal peatland sites and larger areas of pasture zones are indeed degrading, reflecting both local testimonies and satellite image analysis. Further investigation is needed to identify what role community management practices such as irrigation and interannual, long-term environmental shifts, perhaps owing to climate change or hydrological changes, are contributing to the current state and future sustainability of peatlands. The community members were absolutely correct in their assessment that colpares were increasing, and the results of the image analysis show an even greater change in recent decades than anticipated. Salinization processes are occurring in the park and have resulted in an alarming increase in colpares over recent years, driven in part by overgrazing, decrease in pasture management, and possible decrease in water availability in some areas due to climate change.

For several millennia bofedales have been managed by Andean communities to support local livelihoods. However, increasing competition over water resources and climate change patterns may affect the sustainability and resilience of bofedales in coming decades. Continued research on the health of peatlands, and specifically on defining their relationship with hydrological systems, glacier recession, and pasture management practices, is a priority. The strategic conservation of peatland systems will contribute to healthier mountain ecosystems, local carbon sequestration, and the securing of enough pastures for future generations. In the coming decades of rapid and unprecedented social and environmental change peatland health is of paramount importance to sustaining the long Andean tradition of pastoralism.

REFERENCES CITED

Acarapi-Cruz, J. L. 2007. *Análisis de Costos de Producción de Fibra y Carne de Camelidos en el Municipio Curahuara de Carangas.* La Paz: Universidad Mayor de San Andres.

Alzérreca, A. H. 2001. *Los Campos Naturales de Pastoreo del Parque Nacional Sajama (PNS) y su Capacidad de Carga.* La Paz: SERNAP.

Alzérreca, A. H., J. Laura, F. Loza, D. Luna, and J. Ortega. 2006. "Importance of Carrying Capacity in Sustainable Management of Key Andean Puna Rangelands in Ulla Ulla, Bolivia." In *Land Use Change and Mountain Biodiversity*, ed. E. Spehn, M. Liberman-Cruz, and C. Körner, 167–85. Boca Raton: Taylor and Francis.

Baied, C., and J. Wheeler. 1993. "Evolution of High Andean Puna Ecosystems: Environment, Climate, and Cultural Change Over the Last 12,000 Years in the Central Andes." *Mountain Research and Development* 13: 145–56.

Beck, S., A. Domic, C. Garcia, R. I. Meneses, K. Yager, and S. Halloy. 2010. *El Parque Nacional Sajama y sus Plantas.* La Paz: Fundacion PUMA.

Becker, C. D., and K. Ghimire. 2003. "Synergy Between Traditional Ecological Knowledge and Conservation Science Supports Forest Preservation in Ecuador." *Conservation Ecology* 8, no. 1: 1.

Berkes, F. 1999. *Sacred Ecology: Traditional Ecological Knowledge and Management Systems.* Philadelphia: Taylor and Francis.

Bosman, A. F., P. C. van der Molen, R. Young, and A. M. Cleef. 1993. "Ecology of a *Paramo* Cushion Mire." *Vegetation Science* 4: 633–40.

Bradley, R. S., M. Vuille, H. F. Diaz, and W. Vergara. 2006. "Threats to Water Supplies in the Tropical Andes." *Science* 312: 1755–56.

Bradley, R. S., F. T. Keimig, H. F. Diaz, and D. R. Hardy. 2009. "Recent Changes in Freezing Level Heights in the Tropics with Implications for the Deglacierization of High Mountain Regions." *Geophysical Research Letters* 36, no. 17: 4pp.

Browman, D. L., ed. 1987. *Arid Land Use Strategies and Risk Management in the Andes: A Regional Anthropological Perspective.* Boulder: Westview Press.

———. 1989. "Origins and Development of Andean Pastoralism: An Overview of the Past 6000 Years." In *The Walking Larder: Patterns of Domestication, Pastoralism, and Predation,* ed. J. Clutton-Brock, 257–68. London: Unwin Hyman.

Buttolph, L. P. 1998. "Rangeland Dynamics and Pastoral Development in the High Andes: The Camelid Herders of Cosapa, Bolivia." PhD diss., Utah State University.

Buttolph, L. P., and D. L. Coppock. 2004. "Influence of Deferred Grazing on Vegetation Dynamics and Livestock Productivity in an Andean Pastoral System." *Journal of Applied Ecology* 41, no. 4: 664–74.

Campos, M. A., and L. F. Yancas. 2009. *Guía Descriptiva de los Sistemas Vegetacionales Azonales Hídricos Terrestres de la Ecorregión Altiplánica (SVAHT).* Santiago: Ministerio de Agricultura de Chile, Servicio Agrícola y Ganadero.

Cardozo, A., M. d. O. Ismael, L. A. Rodrigo, A. S. Muñoz, and J. L. Tellería, eds. 2004. *Procesos de Salinización en al Altiplano Central: Una Contribución a su Conocimiento.* La Paz: Academia Nacional de Ciencias de Bolivia.

Carey, M. 2008. "The Politics of Place: Inhabiting and Defending Glacier Hazard Zones in Peru's Cordillera Blanca." In *Darkening Peaks: Glacier Retreat, Science, and Society,* ed. B. Orlove, E. Wiegandt, and B. H. Luckman, 229–40. Berkeley: University of California Press.

Chalmers, N., and C. Fabricius. 2007. "Expert and Generalist Local Knowledge about Land-cover Change on South Africa's Wild Coast: Can Local Ecological Knowledge Add Value to Science?" *Ecology and Society* 21, no. 1: 10pp.

Chander, G., and B. Markham. 2003. "Revised Landsat-5 TM Radiometric Calibration Procedures and Postcalibration Dynamic Ranges." *IEEE Transactions on Geoscience and Remote Sensing* 41: 2674–77.

Chimner, R. A., and J. M. Karberg. 2008. "Long-Term Carbon Accumulation in Two Tropical Mountain Peatlands, Andes Mountains, Ecuador." *Mires and Peat* 3, no. 4: 10pp.

Coppock, D. L., and C. Valdivia, eds. 2001. *Sustaining Agropastoralism on the Bolivian Altiplano: The Case of San José Llanga.* Logan: Department of Rangeland Resources, Utah State University.

Cooper, D. J., E. C. Wolf, C. Colson, W. Vering, A. Granda, and M. Meyer. 2010. "Alpine Peatlands of the Andes, Cajamarca, Peru." *Arctic, Antarctic, and Alpine Research* 42, no. 1: 19–33.

Cooper, D., K. Kaczynski, D. Slayback, K. Yager. "Growth, Production, and Short-Term Peat Accumulation in Distichia Muscoides Dominated Bofedales, Bolivia, South America." *Plant Ecology,* submitted.

Coudrain, A., B. Francou, and Z. W. Kundzewicz. 2005. "Glacier Shrinkage in the Andes and Consequences for Water Resources." *Hydrological Sciences Journal–Journal des Sciences Hydrologiques* 50, no. 6: 925–32.

Diaz, H., and N. Graham. 1996. "Recent Changes in Tropical Freezing Heights and the Role of Sea Surface Temperature." *Nature* 383: 152–55.

Diaz, H., J. K. Eischeid, C. Duncan, and R. S. Bradley. 2003. "Variability of Freezing Levels, Melting Season Indicators, and Snow Cover for Selected High-Elevation and Continental Regions in the Last 50 Years." *Climatic Change* 59: 33–52.

Dise, N. 2009. "Peatland Response to Global Change." *Science* 326: 810–11.

Earle, L. R., B. G. Warner, and R. Aravena. 2003. "Rapid Development of an Unusual Peat-Accumulating Ecosystem in the Chilean Altiplano." *Quaternary Research* 59: 2–11.

Espinoza, C. W. 2001. *Caracterización de los Sistemas de Producción del Parque Nacional Sajama y sus Zonas de Amortiguación.* La Paz: MAPZA-GTZ-PNS-SERNAP.

Flores-Ochoa, J. 1977. *Pastores de Puna: Uywamichiq Punarunakuna.* Lima: Instituto de Estudios Peruanos.

Francou, B., E. Ramirez, B, Cáceres, and J. Mendoza. 2000. "Glacier Evolution in the Tropical Andes During the Last Decades of the 20th Century: Chacaltaya, Bolivia, and Antizana, Ecuador." *Ambio* 29, no. 7: 416–22.

Gadgil, M., P. Olsson, F. Berkes, and C. Folke. 2003. "Exploring the Role of Local Ecological Knowledge for Ecosystem Management: Three Case Studies." In *Navigating Social-Ecological Systems: Building Resilience for Complexity and Change,* ed. F. Berkes, J. Colding, and C. Folke, 189–209. Cambridge: Cambridge University Press.

Georges, C. 2004. "The 20th Century Glacier Fluctuations in the Tropical Cordillera Blanca, Peru." *Arctic, Antarctic and Alpine Research* 36, no. 1: 100–107.

Gorham, E. 1991. "Northern Peatlands: Role in the Carbon Cycle and Probable Response to Climatic Warming." *Ecological Applications* 1, no. 2: 182–95.

Hardy, D. R., M. Vuille, and R. Bradley. 2003. "Variability of Snow Accumulation and Isotopic Composition on Nevado Sajama, Bolivia." *Journal of Geophysical Research* 108: 46–93.

Hardy, D. R., M. Vuille, C. Braun, F. Keimig, and R. Bradley. 1998. "Annual and Daily Meterological Cycles at High Altitude on a Tropical Mountain." *Bulletin of the American Meteorological Society,* 1899–1913.

Hoffmann, D. 2007. "The Sajama National Park in Bolivia: A Model for Cooperation Among State and Local Authorities and the Indigenous Population." *Mountain Research and Development* 27, no. 1: 11–14.

IPCC (Intergovernmental Panel on Climate Change). 2007. "Climate Change 2007: Impacts, Adaptation and Vulnerability—Summary for Policymakers." Geneva: Intergovernmental Panel on Climate Change.

Kaser, G. 1999. "A Review of the Modern Fluctuations of Tropical Glaciers." *Global and Planetary Changes* 22: 93–103.

Liberman-Cruz, M. 1986. "Microclima y Distribución de Polylepis Tarapacana en el Parque Nacional del Nevado Sajama, Bolivia." *Documents Phytosociologiques* 10, no. 2: 235–72.

MAPZA 2001. (Proyecto Manejo de Areas Protegidas y Zonas de Amortiguación). "Caracterización de los Sistemas de Producción del Parque Nacional Sajama a Partir de los Estudios de Caso y la Caracterización Comunal." La Paz: SERNAP, MAPZA-GTZ.

Mark, B. G., and G. O. Seltzer. 2003. "Tropical Glacier Meltwater Contribution to Stream Discharge: A Case Study in the Cordillera Blanca, Peru." *Journal of Glaciology* 49: 271–81.

Maydana, D. 2010. "El Rencuentro de la Vicuña con las Comunidades Aymaras." *Grupo Especialista en Camélidos Sudamericanos.*

Mayer, E. 2001. *The Articulated Peasant: Household Economies in the Andes.* New Haven: Yale University Press.

Mohr, K., D. Slayback, and K. Yager. 2014. "Characteristics of Precipitation Features and Annual Rainfall During the TRMM Era in the Central Andes." *Journal of Climate* 27: 3982–4001.

Moore, P. D. 1989. "Ecology of Peat-Forming Processes: A Review." *International Journal of Coal Geology* 12: 89–103.

Murra, J. 1965. "Herds and Herders in the Inca State." In *Man, Culture and Animals*, ed. A. Leeds and A. Vayda, 185–215. Washington, D.C.: American Association for the Advancement of Science.

Orlove, B. 1977. *Alpacas, Sheep and Men: The Wool Export Economy and Regional Society of Southern Perú.* New York: Academic Press.

Orlove, B., J. Chiang, and M. Cane. 2000. "Forecasting Andean Rainfall and Crop Yield from the Influence of El Niño on Pleiades Visibility." *Nature* 403: 68–71.

Palacios-Ríos, F. 1992. "Pastizales de Regadío para Alpacas en la Puna Alta (el Ejemplo de Chichillapi)." In *Comprender la Agricultura Campesina en los Andes Centrales, Perú y Bolivia*, ed. P. Morlon, 207–13. Lima: Instituto Francés de Estudios Andinos and Centro de Estudios Regionales Andinos Bartolomé de las Casas.

Patty, L., S. Halloy, E. Hiltbrunner, and C. Körner. 2010. "Biomass Allocation in Herbaceous Plants Under Grazing Impact in the High Semi-Arid Andes." *Flora* 205: 695–703.

Postigo, J. C., K. R. Young, and K. A. Crews. 2008. "Change and Continuity in a Pastoralist Community in the High Peruvian Andes." *Human Ecology* 36: 535–51.

Rabatel, A., B. Francou, A. Soruco, J. Gomez, B. Caceres, J. L. Ceballos, R. Basantes, M. Vuille, J. E. Sicart, C. Huggel, M. Scheel, Y. Lejeune, Y. Arnaud, M. Collet, T. Condom, G. Consoli, V. Favier, V. Jomelli, R. Galarraga, P. Ginot, L. Maisincho, J. Mendoza, M. Menegoz, E. Ramirez, P. Ribstein, W. Suarez, M. Villacis, and P. Wagnon. 2013. "Current State of Glaciers in the Tropical Andes: A Multi-Century Perspective on Glacier Evolution and Climate Change." *Cryosphere* 7: 81–102.

Rhoades, R., X. Z. Rios, and J. Arangundy. 2006. "Climate Change in Cotacachi." In *Development with Identity: Community, Culture and Sustainability in the Andes*, ed. R. E. Rhoades, 64–74. Wallingford, UK: CAB International.

Rocha O., and C. Sáez, eds. 2003. *Uso Pastoril en Humedales Altoandinos: Talleres de Capacitación para el Manejo Integrado de los Humedales Altoandinos de Argentina, Bolivia, Chile y Perú.* La Paz: Convención RAMSAR, WCS-Bolivia.

Roulet, N., T. Moore, J. Bubier, and P. Lafleur. 1992. "Northern Fens: Methane Flux and Climate Change." *Tellus* 44B: 100–105.

Ruthsatz, B. 1993. "Flora and Ecological Conditions of High Andean Peatlands of Chile Between 18°00' (Arica) and 40°30' (Osorno) South Latitude." *Phytoenologia* 25: 185–234.

———. 2000. "Die Hartpolstermoore der Hochanden und ihre Artenvielfalt. *Ber. D. Reinh. Tuexen-Ges. (Hannover)* 12: 351–71.

San Martin, F., and F. C. Bryant. 1987. "Nutrición de los Camélidos Sudamericanos: Estado de Nuestro Conocimiento." Lubbock: Texas Tech University Technical Article.

Seimon, T. A., A. Seimon, P. Daszak, S. Halloy, L. Schloegel, C. Aguilar, P. Sowell, A. Hyatt, B. Konecky, and J. Simmons. 2007. "Upward Range Extension of Andean Anurans and *Chytridiomycosis* to Extreme Elevations in Response to Tropical Deglaciation." *Global Change Biology* 13: 288–99.

Silverio, W., and J. M. Jaquet. 2005. "Glacial Cover Mapping (1987–1996) of the Cordillera Blanca (Peru) Using Satellite Imagery." *Remote Sensing of the Environment* 95: 342–50.

Slayback, D., K. Yager, M. Baraer, K. Mohr, J. Argollo, O. Wigmore, R. Meneses, and B. Mark. 2012. "Changing Hydrology in Glacier-fed High Altitude Peatlands." American Geophysical Union Conference, San Francisco.

Soruco, A., C. Vincent, B. Francou, and J. Francisco Gonzales. 2009. "Glacier Decline Between 1963 and 2006 in the Cordillera Real, Bolivia." *Geophysical Research Letters* 36, no. L03502: 6.

Squeo, F. A., B. G. Warner, R. Aravena, and D. Espinoza. 2006. "Bofedales: High-Altitude Peatlands of the Central Andes." *Revista Chilena de Historia Natural* 79: 245–55.

Strack, M. 2008. *Peatlands and Climate Change.* Jyväskylä, Finland: International Peat Society.

Thompson, L. G., M. E. Davis, E. Mosley-Thompson, T. A. Sowers, K. A. Henderson, V. S. Zagorodnov, P. N. Lin, V. N. Mikhalenko, R. K. Campen, J. F. Bolzan, J. Cole-Dai, and B. Francou. 1998. "A 25,000-Year Tropical Climate History from Bolivian Ice Cores." *Science* 282, no. 5295: 1858–64.

Tolenen, K., and J. Turunen. 1996. "Accumulation Rate of Carbon in Mires in Finland and Implications for Climate Change." *Holocene* 6: 171–78.

Troll, C. 1968. "The Cordilleras of the Tropical Americas: Aspects of Climatic, Phytogeographical and Agrarian Ecology." In *Geo-ecology of the Mountainous Regions of the Tropical Americas,* ed. C. Troll, 15–56. Bonn: Dümmlers Verlag.

Tucker, C. 1986. "Maximum Normalized Difference Vegetation Index for Images for Sub-Saharan Africa for 1983–1985." *International Journal of Remote Sensing* 7: 1383–84.

Ulloa, D., and K. Yager. 2007. *Memorias del Taller "Cambio Climático: Percepción Local y Adaptaciones en el Parque Nacional Sajama."* La Paz: Conservation International–Bolivia.

UNEPCA (Proyecto de Desarrollo de Criadores de Camélidos del Altiplano Boliviano). 1999. *Censo Nacional Bolivia de Llamas y Alpacas.* La Paz: Centro de Información para el Desarrollo (CID).

Urrutia, R., and M. Vuille. 2009. "Climate Change Projections for the Tropical Andes Using a Regional Climate Model: Temperature and Precipitation Simulations for the End of the 21st Century." *Journal of Geophysical Research–Atmospheres* 114: doi:10.1029/2008JD011021.

Vergara, W., A. M. Deeb, A. M. Valencia, R. S. Bradley, B. Francou, A. Zarzar, A. Grunwaldt, and S. M. Haeussling. 2007. "Economic Impacts of Rapid Glacier Retreat in the Andes." *Eos, Transactions American Geophysical Union* 88, no. 25: 261–68.

Vuille, M. 1999. "Atmospheric Circulation over the Bolivian Altiplano During Dry and Wet Periods and Extreme Phases of the Southern Oscillation." *International Journal of Climatology* 19: 1579–1600.

———. 2007. *Climate Change in the Tropical Andes: Impacts and Consequences for Glaciation and Water Resources. Part 1: The Scientific Basis.* Amherst: A Report for CONAM and the World Bank.

Vuille, M., and R. Bradley. 2000. "Mean Annual Temperature Trends and Their Vertical Structure in the Tropical Andes." *Geophysical Research Letters* 27: 3885–88.

Vuille, M., R. S. Bradley, M. Werner, and F. Keimig. 2003. "20th Century Climate Change in the Tropical Andes: Observations and Model Results." *Climatic Change* 59: 75–99.

Vuille, M., B. Francou, P. Wagnon, I. Juen, G. Kaser, B. G. Mark, and R. S. Bradley. 2008. "Climate Change and Tropical Andean Glaciers: Past, Present and Future." *Earth-Science Reviews* 89: 79–96.

Vuille, M., D. R. Hardy, C. Braun, F. Keimig, and R. Bradley. 1998. "Atmospheric Circulation Anomalies Associated with 1996/1997 Summer Precipitation Events on Sajama Ice Cap, Bolivia." *Journal of Geophysical Research* 103, no. D10: 11191–11204.

Washington-Allen, R. A., R. D. Ramsey, B. E. Norton, and N. E. West. 1998. "Change Detection of the Effect of Severe Drought on Subsistence Agropastoral Communities on the Bolivian Altiplano." *International Journal of Remote Sensing* 19: 1319–33.

Yager, K. 2009. "A Herder's Landscape: Deglaciation, Desiccation, and Managing Green Pastures in the Andean Puna." PhD diss., Yale University.

Yager, K., H. Resnikowski, and S. Halloy. 2008a. "Grazing and Climatic Variability in Sajama National Park, Bolivia." *Pirineos* 163: 97–109.

Yager, K., D. Ulloa, and S. Halloy. 2008b. "Conducting an Interdisciplinary Workshop on Climate Change: Facilitating Awareness and Adaptation in Sajama National Park, Bolivia." In *Interdisciplinary Aspects of Climate Change,* ed. W. L. Filho, 327–42. Hamburg: Hamburg University of Applied Sciences.

Young, K., and J. Lipton. 2006. "Adaptive Governance and Climate Change in the Tropical Highlands of Western South America." *Climatic Change* 78: 63–102.

Chapter 7 **Challenges in Integrating the Climate and Social Sciences for Studies of Climate Change Impacts and Adaptation**

Frances C. Moore, Justin S. Mankin, and Austin Becker

The study of climate change has grown rapidly in both size and scope over the past forty years. This growth has coincided with an expansion in the breadth of climate change science as the community seeks not just to improve projections of temperature, precipitation, and sea level rise, but also to trace the consequences of these physical changes on social, economic, and natural systems through climate change impacts research. It also corresponds to growth in the scope of climate policy; the set of legitimate responses to the climate change problem has grown from energy policy alone to include forestry and agriculture and, under the term *adaptation*, adjustments in all systems impacted by climate change.

Despite this growth, challenges remain in reconciling the contextualized and interpretive social sciences, such as anthropology, with the climate sciences. We address why this might be the case, focusing on studies of climate change impacts and adaptation in social and economic systems. This area of climate change research is one in which integrated, cross-disciplinary analysis is particularly needed, as it speaks to the ways in which humans perceive and respond to their environment. Natural and social scientists have done important work in this area, but too often these rich and nuanced areas of research struggle to be substantively and meaningfully integrated.

To explore why this is so, we draw on our experiences working at the interface between the climate and social sciences in an interdisciplinary doctoral program. We pursue work on climate change impacts and adaptation in different sectors, but we all have both natural and social scientists on our dissertation committees. The common difficulties we have encountered, both in executing the research itself and in justifying the same piece of work to faculty from very different methodological and epistemological traditions, inform the following discussion on why fuller integration remains a challenge. Below we outline three specific areas in which these traditions are at odds: (1) the linked question of temporal and spatial scale of analysis, (2) the need to isolate climate effects from other effects, and (3) the differing attitudes toward prediction. These considerations are followed by a discussion that includes the receptivity of the climate science community to social science inputs and suggestions for some promising ways forward.

SCALE

Climate scientists tend to work at large spatial scales and long timescales. In this section, we describe how this is a result of the computational limitations of climate models and the types of uncertainty involved and explore the associated difficulties for integration with social sciences. Although climate science involves a range of data and methods, the use of large, computationally intensive General Circulation Models (GCMs) to model the climate system and project the effects of greenhouse gas emissions dominates the field and is our focus here. As background it is helpful to understand three types of uncertainties in climate model projections (Hawkins and Sutton 2009):

1. Internal variability is the natural fluctuations in weather resulting from the nonlinear, chaotic nature of the climate system that occur even in the absence of any forcing from human-induced climate change.
2. Model uncertainty is the difference in model responses to the same amount of climate change and can be thought of broadly as our scientific uncertainty over how the climate system operates and responds to greenhouse gas emissions.

3. Scenario uncertainty results from the fact that we do not know how technology, policy, and the economy will shift over time and therefore how emissions will change in the future.

Temporal Scale

Climate scientists are most comfortable using very long timescales. They are typically uncomfortable looking at data at timescales of less than multiple decades, and often results from climate models are reported for fifty or one hundred years into the future. Such insistence on very long-term analysis can run counter to the social sciences, where timescales might more usually be defined by seasons, business planning horizons, political election cycles, or human lifetimes.

Climate is a phenomenon defined statistically over long timescales. The IPCC Fourth Assessment Report uses the following description: "The atmospheric component of the climate system most obviously characterizes climate; climate is often defined as 'average weather.' Climate is usually described in terms of the mean and variability of temperature, precipitation and wind over a period of time, ranging from months to millions of years (the classical period is 30 years)" (IPCC 2007b: 96). This definition emphasizes the long, though flexible, temporal scale over which climate is defined and its statistical nature. A physicist once made this helpful distinction: "We dress in the morning for weather, whereas we build a house for climate."[1]

The long timescales at which climate scientists tend to operate can sometimes seem to stretch commonly understood meanings of words. For example, "fast feedbacks" in the climate system are those processes (such as additional water vapor in the atmosphere from higher temperatures) that reinforce a climatic trend on timescales anything shorter than a decade (Hansen et al. 1984; Lorius et al. 1990). For paleoclimatologists studying Earth's past climates as possible analogs for a warmer planet, *fast* can mean anything from hundreds to tens of thousands of years.

Looking backward, climate scientists' preference for long timescales of analysis stems from the variability inherent to the earth systems that affect climate and the desire to characterize an accurate normal baseline against which anthropogenic changes can be measured (the "internal variability" type of uncertainty described above). Only over long periods of time can climate scientists observe the most important internal cycles and define a baseline measure from which to discern the signal of a changing climate against the statistical noise of natural climate variability. Given even decades of observations in the modern record, debate exists as to whether trends in certain observations, North Atlantic hurricane intensity, for example, are abnormal exceptions attributable to anthropogenic climate change or part of natural long-term internal oscillations of the Earth's climate system (IPCC 2012: 159–61).

Looking forward, long timescales are equally important. Climate phenomena take time to manifest because it takes time for energy to be redistributed around the planet through the ocean. This "thermal inertia," as the slow redistribution of excess energy throughout the Earth is called, slows the temperature response of the Earth because there is energy in the system that is not yet realized as atmospheric warming (Hansen et al. 1985, 2005). This feature of the climate system has two implications. First it adds a degree of determinism and predictability. In the mean, over a long period of time, this energy will manifest in the climate system, which gives rise to weather that people experience. Second, such temporal lags imply that the full warming from the current concentrations of greenhouse gases has yet to be realized. There is a severe disconnect between the emission of greenhouse gases by humans today and the timescales on which the consequences of those actions will be realized. This timescale is determined by a combination of climate sensitivity and the rate at which the ocean mixes (Hansen et al. 1985, 2005).

These are important reasons why climate scientists tend to be most comfortable operating at timescales of at least several decades or more. However, studying the human consequences of climate change impacts over such long timeframes is complicated. Looking backward, huge social changes have occurred in the past fifty to one hundred years, making it extremely difficult to distinguish what effect, if any, climate change has had in that time period. Correspondingly, over that same time period in the future, similarly large and unforeseeable changes make it difficult to know even where to begin in predicting the effects of climate change one hundred years from now.

This effect is clear when uncertainty over climate projections is divided into the three types described above, as shown in Hawkins and Sutton (2009, reproduced as fig. 7.1). At timescales of less than fifty years, natural variability and model differences dominate the uncertainty in projections, but over longer time horizons, scenario uncertainty comes to dominate, accounting for over 80 percent of total projection uncertainty by 2100. In other words, projections of the effect of a given level of greenhouse gas

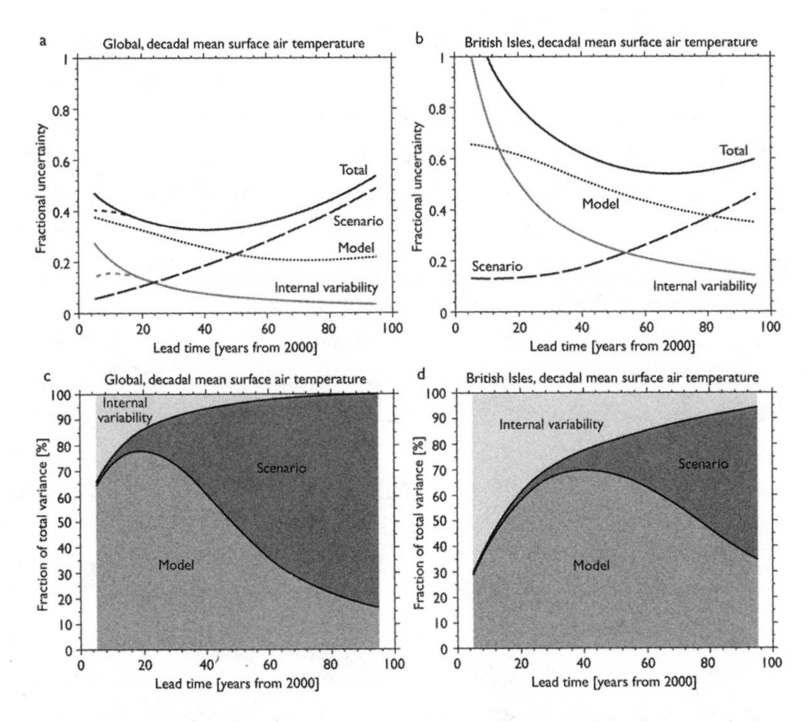

7.1: Sources of variance in climate model predictions. Hawkins and Sutton (2009: 1097)

emissions become more certain further into the future, but uncertainty over what that level will be (and, relatedly, the effect it will have on society and the economy) become much less certain far into the future. One might say that the climate-science component of uncertainty decreases with time but that the social-science component (imperfectly captured here as scenario uncertainty) increases with time. Given this difference, it is perhaps unsurprising that the "natural" timescale of impact analysis should be very different in the climate and social sciences.

Spatial Scale

Spatial scale presents another set of challenges for those trying to integrate climate modeling with its social, economic, and political implications. Decision making that impacts us most immediately and directly tends to be understood on a local or perhaps regional scale: How much will sea level rise in my coastal community? Will I need to shift to a new, more drought-resistant crop? Ultimately, this is the scale at which people, and therefore often social scientists, engage with the issue of climate change impacts. And yet there are huge methodological challenges associated with studying climate change at anything less than continental scales.

Climate change is often called the quintessential global environmental problem. It is perhaps the typical example of a broader shift in the scale of environmental concerns from local to global since the foundation of the global environmental governance regime in 1972 and the start of satellite imaging and global data-monitoring efforts (Forsyth 2003: 168–201). The typical reason given for this global characterization focuses on causes: because atmospheric CO_2 is well mixed in the atmosphere, emissions in one location affect every other location on the globe. However, the geographic interchangeability of CO_2 emissions and political treatment in large, multinational fora is only one reason for the satellite-eye view so pervasive in climate science today. Another important reason prominent in the climate change impacts literature is that, for any given timescale, climate change impacts are most easily observed and most easily predicted at larger spatial scales.

Observing climate changes realized to date and establishing the cause of such changes constitutes a subfield of climate science known as detection and attribution. This work was first applied to more basic physical climate changes such as temperature (Tett et al. 1999; Thorne et al. 2003), sea level rise (Church and White 2006) and precipitation patterns (Zhang et al. 2007). Because the climate change that has occurred to date is a small but widely diffused effect, detection efforts can work only at the largest spatial scales, that is subcontinental, continental, hemispheric, and global).[2] At this scale, the internal variability of systems and any changes owing to nonclimatic factors can be expected to cancel each other out, allowing the weak climate change signal to be distinguished from large amplitude noise. At smaller scales, the impacts of climate change to date tend to be dwarfed by particular local factors. The example of sea level rise is illustrative. Climate change is thought to have raised global sea level at an accelerating rate over the past century, but tidal gauge stations in any one location may be influenced by a number of local factors, including regional tectonic activity related to the last ice age, groundwater withdrawal, dense urbanization, or coastal sedimentary processes. Only in the aggregate can these records produce compelling evidence that the sea level rise expected from climate change has in fact occurred. It is this need for long-term, large-scale data series that has made the climate sciences such voracious consumers of all kinds of long-term data from around the world (Edwards 2010: 287–336).[3]

More recently, detection and attribution efforts have moved into second-order climate change effects more directly tied to social, economic, and natural impacts such as shifts in phenology and species ranges (Parmesan and Yohe 2003; Root et al. 2003), snowpack cover (Pierce et al. 2008), agricultural seasons (Estrella et al. 2007), and stream flow (Barnett et al. 2008). Identifying the effects of climate change on more socially relevant but far less proximate areas such as livelihoods, human health, migration patterns, food security, and conflict requires tracing the climate change signal through ever more complex and noisy layers of social, economic, and political interactions.

Traditionally, detection and attribution have been done statistically, over large spatial scales. The reason for this is that it is not enough to simply document that a change has taken place (detection); one must also demonstrate that the pattern of this change is consistent only with anthropogenic climate change rather than with other possible causes (attribution). For example, looking at the pattern of temperature trends over the whole atmosphere and comparing these with trends that would be expected from other causal mechanisms (generated by using model simulations), scientists can show that the pattern of observed warming closely matches the "fingerprint" that would be expected from human-induced greenhouse gas emissions and is inconsistent with other causes such as aerosols or changing solar activity (Santer et al. 1996).

Looking forward, projections of the future impacts of climate change tend to be most accurate when giving average predictions at larger spatial scales. Partly this relates to the low resolution of global climate models, typically run at grid scales of 150km or greater. Any processes important in driving weather or climate variability at smaller scales remain unaccounted for or roughly parameterized. Computational cost, or the amount of time it takes for a climate model to calculate its answers, increases rapidly with spatial resolution. For example, the new supercomputer at the National Center for Atmospheric Research, which has a resolution of about 25 km^2, is able to simulate 1 year per 24 real-world hours.[4] This gives a century-long simulation a computational cost of over three months. Thus computational resources are a major constraint on the spatial resolution of climate model projections.

In addition, in the medium term (the next two to three decades) most relevant for projections of social and economic impacts, the climate signal might be weak compared to the internal variability at any one location. Climate models can only roughly characterize this internal variability because computational resources limit the number of repeat runs that can be done for a given simulation. This internal variability is often, though not uniformly, smaller when averaged over larger spatial scales. Deser et al. (2012) look at how this internal variability affects the range of possible future warming trends over the next fifty years at different spatial scales by comparing forty runs of the same model using the same emissions scenario differing only in their realizations of internal variability. These fifty-year trends at the global scale show almost no differences between the model runs, but at the scale of North America, predicted temperature changes ranged from less than 1 degree Celsius to almost 3.5 degrees. Looking at smaller scales, predicted temperature changes for Seattle, Washington, varied even more, from a projected cooling of 0.1 degrees to a warming of over 4 degrees.

Therefore, looking both backward and forward, the typical scale for climate change impact analysis tends to be large, if not global, because errors associated with both subgrid scale processes and internal variability are more likely to cancel at larger spatial scales. Certainly the spatial scale is much larger than the typical scale of any of the social sciences. Demand for impact studies at scales relevant for planning as well as the availability of higher-resolution regional climate models has begun to shrink the scale of analysis in recent years, but the law of averages remains. Conclusions regarding climate change impacts will always be more certain at larger spatial scales, exerting an upward push in the scale at which climate scientists are most comfortable operating. This is in stark contrast to contextualized social sciences, which often focus at smaller spatial scales.

The previous two sections have argued that climate scientists work at both longer timescales and larger spatial scales than social scientists as a way of managing uncertainties due to natural variability and climate model differences. But the decomposition of uncertainty in future temperature projections by Hawkins and Sutton (2009) shows that space and time are related and somewhat substitutable in this respect. The uncertainty in global temperature projections is minimized approximately forty years in the future (see fig. 7.1a)—meaning scientists are most confident in projections at this timescale. Forty years is early enough that the set of possible futures captured by emissions scenarios has not diverged substantially (scenario uncertainty is still low), but far enough for the climate change signal to be large compared to both the internal variability in the climate system and model uncertainty. This point is sometimes informally referred to as the sweet spot by the climate modeling community. However, the same decomposition for projections of mean temperature over the British Isles (see fig. 7.1b) shows a much later sweet spot between sixty and seventy years in the future. This is because larger internal and model variability at the smaller spatial scale means it takes longer for the anthropogenic signal of climate change to become discernible. Thus a smaller spatial scale of analysis can sometimes be compensated for by a longer timescale, but it will rarely make sense, from a climate modeler's perspective, to study local climate over short (less than thirty years) timescales. Unfortunately, this is precisely the scale at which demand for climate change projections tends to be greatest.

ABSTRACTING CLIMATE-RELATED CHANGE FROM OTHER CHANGES

Climate change impact studies, by definition, focus on changes attributable to climate change. Most commonly, results are reported as changes in some outcome of interest relative to what would have happened in the absence of climate change. Such statements require developing a baseline point of comparison, and these hypothetical, ultimately unobservable estimated trajectories pervade the climate science and policy literatures. Developing these baselines involves subjective and sometimes opaque or unjustified assumptions that may dramatically change results (see also Matthews, this volume).

Moreover, such baselines require exponentially more assumptions for impact assessments further down the "impact chain" that connects anthropogenic CO_2 emissions to changes in social, economic, and natural systems of interest. Food security analyses are a good example. Though estimating changes in temperature and precipitation patterns attributable to anthropogenic climate change is not trivial, a large, established modeling community exists to do exactly that, through well-established best practices that make the assumptions of this process both transparent and fairly common across modeling groups. Translating these changes into projections of agricultural yields requires not just crop-physiology models but also estimates of the rate of yield increase due to agronomic research, changes in the use of fertilizer and pesticide, and changes in irrigation practices. Evaluating the ultimate outcome of interest, food security, requires even more assumptions regarding biofuel policy, global trade practices, population growth, agricultural subsidies, demand elasticity, urbanization, income growth and its effect on dietary preferences, the ease with which additional land can be brought under cultivation, and the ability of farmers to offset production declines through adaptation (Hertel 2010; Hertel et al. 2010; Rosenzweig and Parry 1994).

Understandably, few researchers can undertake to thoroughly project all of these, so impact studies typically model only a few while making simplifying assumptions or assuming no change for the remainder. The use of up to three or four coupled models, each with particular sets of assumptions, and the lack of standards of practice in the impacts modeling community regarding baseline development can make understanding these assumptions and how they might shape results difficult and time consuming. But this model-cascade methodology is needed for exploring hypothetical counterfactuals if the goal of research is to isolate, quantify, and ultimately predict the effect of climate change on some outcome of interest.

However, a more integrated perspective from the social sciences brings into question whether such reductionist inquiries are possible or even desirable. Any researcher seeking to identify climate change impacts in the context of the lived experience of people affected will quickly run into the question, Why climate? Other risks, uncertainties, and ongoing social, political, and economic changes almost certainly dominate decisions in daily life, so that a single-minded focus on the climate increment can seem pedantic and academic. This is not to say that climate change impacts are insignificant—even relatively small, gradual impacts can have a large effect if widely distributed through the globalized economy.

But attempting to engage with subjects about climate change can be difficult when it seems so marginal to other forces shaping their lives (Barnes, this volume). This is true even when talking to people tasked with making decisions involving long-lived infrastructure directly impacted by climate change. For example, decision makers who make plans and investments in long-lasting infrastructure like seaports base their decisions on economic models that incorporate return on investment over the economic life of the project. A given wharf, pier, or terminal is designed to pay for itself over the term of its loans. However, structures built today can reasonably be expected to continue performing far beyond this thirty- to fifty-year time horizon. With sea level rise, the failure rate of storm surge designs will almost certainly be higher by the end of the century, but few port decision makers are actively considering sea level rise projections in planning in favor of shorter-term budgetary concerns (Becker et al. 2011).

This disconnect has given rise to a divide in the climate change impacts community along rough disciplinary lines, which O'Brien et al. (2004) refer to as the "end-point vulnerability" and "starting-point vulnerability" perspectives. The former perspective, common among natural scientists and economists, defines climate change vulnerability as the residual effect of climate change after adaptation. Methodologically, this perspective emphasizes quantitative prediction of climate change impacts on social and economic systems (Diffenbaugh et al. 2011; Luers et al. 2003; Nelson et al. 2010; Reidsma et al. 2009). This can be done either by using models that simulate the system to look at outcomes with and without climate change ("process-based models") or by looking at statistical relationships between climate and outcome to project the marginal effect of a changing climate. This approach was prominent in the IPCC Third Assessment Report, in which vulnerability was defined as the exposure of a system, increased by its sensitivity and reduced by adaptive capacity (IPCC 2001: 881).

The starting-point vulnerability perspective is more common in studies informed by anthropology, sociology, or political science and draws heavily on the resilience literature as well as political ecology (Engle 2011). In this framework, the impact assessment process begins by describing vulnerability as a general characteristic of the system of interest, which partly defines the capacity to adapt (O'Brien et al. 2004). The question then becomes to what extent climate change might interact with and exacerbate these existing sources of vulnerability, particularly in relation to other ongoing social changes, and what ability and incentives people have to respond to that change. The goal is generally less quantitative prediction than understanding of the processes and interactions involved, possibly for informing and prioritizing adaptation interventions. This more contextualized, holistic approach to vulnerability was more prominent in the IPCC Fourth Assessment Report: "The vulnerability of a society is influenced by its development path, physical exposures, the distribution of resources, prior stresses, and social and government institutions. . . . Adaptive capacities are unevenly distributed, both across countries and within societies. The poor and marginalized have historically been most at risk, and are most vulnerable to the impacts of climate change" (IPCC 2007: 720). Emphasizing the role of general determinants of vulnerability makes these both the departure point for analysis of climate change impacts and possible points of intervention for adaptation projects. It opens up the conversation from one targeted on specific impact pathways related to climate change to a more general discussion of the role of poverty, inequality, and political marginalization in exacerbating risk, of which climate change is just one of many.

We have formulated an impact-chain framework for understanding these two perspectives on vulnerability and how they interact (fig. 7.2). Any

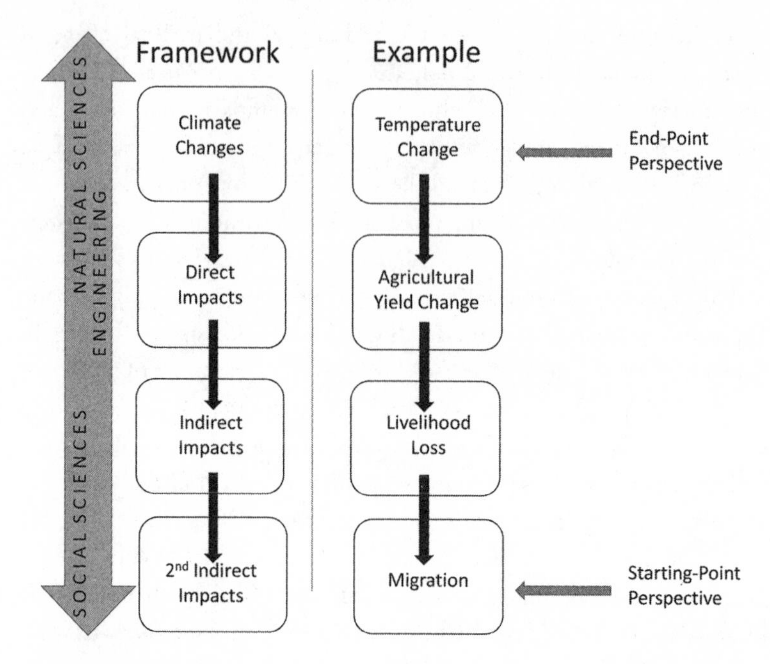

7.2: "Impact-chain" framework. Frances C. Moore, Justin S. Mankin, and Austin Becker.

narrative that connects physical climatic changes to outcomes of interest to society can be broken into a chain of consecutive impacts. In the figure we give the example of higher temperatures leading to yield declines, which in turn lead to livelihood loss and migration. Except for the beginning and end of the chain, it tends to be the linkages between the levels that are the objects of research interest: just how strong is the causal connection between one level and the next? In addition, each linkage offers the opportunity for adaptations to reduce the impact of a given amount of climate change on the ultimate outcome of interest, in this case, migration. For example, changing planting dates could reduce the impact of a given temperature change on yields (first linkage), whereas supplementing household income with off-farm wage labor could reduce the impact of a given yield change on livelihood loss (second linkage).

Different disciplines tend to situate themselves at different points in the chain. The natural sciences typically start with climate change itself and work down by using a sequence of coupled models (an end-point vulnerability perspective), and the social sciences start with indirect and second indirect effects and work up, often qualitatively, to understand their root causes (a starting-point vulnerability perspective). The reason the starting-point perspective leads to a much broader analysis than the end-point perspective is that beginning an analysis with the connection between, for example, livelihood loss and migration makes it clear that climate change may represent only a vanishingly small fraction of this relationship. The end-point perspective instead focuses exclusively on the small part of the connection that can be explained by climate change. One could imagine a huge number of possible causal pathways resulting in livelihood loss and subsequent migration. In the starting-point vulnerability perspective all of these are recognized as being relevant, whereas only the single chain that begins with anthropogenic climate change would be considered relevant in the end-point vulnerability perspective.

Although many people who study climate change impacts, particularly in the IPCC, are trying to integrate analysis of climate change impacts with other risks and vulnerabilities faced by communities in starting-point vulnerability analyses, the impact this will have on policy remains unclear. Most funding sources for climate change adaptation require proof that the project will address only additional burdens imposed by climate change, known as the additionality requirement. Such requirements become absurd in the face of severe financial constraints in developing countries. For example, Tuvalu received funding from the Global Environment Facility for the top half of a seawall to protect against rising seas, but the project was left in limbo in the absence of financing for the bottom half (Ayres and Huq 2009). The recent World Bank estimate of adaptation costs in developing countries is another extreme example of this additionality mindset. This report estimates the "cost of development initiatives needed to restore welfare to levels prevailing before climate change" but insists that such adaptation cost includes "only the additional cost to deal with future climate change" (World Bank 2010: 20, 2). In other words, the bank report envisions a set of technical interventions sufficient to eliminate the net effect of climate change in developing countries but that have no effect on vulnerability to current climatic variability or natural disasters.

Isolating and quantifying the climate change increment through additionality studies is politically desirable because it holds out the promise that the effects of climate change can be addressed without engaging broader issues relating to inequitable wealth, resource use, and vulnerability to natural disasters. It allows climate change policy to remain contained within the specific and technical environmental realm when inquiries into the root causes of climate change vulnerability threaten to spill over into questions of inequality, political marginalization, and inequitable trade relationships. In this respect, climate change adaptation differs little from discourse in other areas of environment and development. Political ecologists have long traced the effect that a technical discourse, which limits the description of environmental problems to their proximate causes and ignores longer-term contexts that include political marginalization and exploitation, has in constraining the set of possible remedies (Blaikie 1985). This process makes complex, highly politicized social problems appear amenable to technical and purportedly apolitical interventions that can be provided by development projects but, because root causes are ignored, may also render these interventions ultimately ineffective (Ferguson 1994; Li 2007).

These insights from political ecology, hard won from decades of past development practice, suggest the climate-change additionality project is impossible. However, end-point vulnerability analyses that quantify and isolate the effect of climate change lend this project a scientific legitimacy by implying that those made hungry or destitute or homeless because of climate change are somehow qualitatively and identifiably distinct from those made hungry or destitute or homeless because of other ongoing social processes. A more integrative starting-point vulnerability perspective threatens this distinction by focusing on common sources of vulnerability, but it remains unclear to what extent it can influence an international policy process that, for better or worse, remains focused on climate change. Both perspectives are well represented within the most recent IPCC Working Group 2 Fifth Assessment Report, but true integration into adaptation theory and policy remains challenging.[5]

ATTITUDE TOWARD PREDICTION

Finally, the climate and social science traditions differ markedly in the level of importance they ascribe to prediction. The climate sciences, particularly today, heavily emphasize predicting the future impacts of climate change, many of which may not become apparent for decades. The policy relevance of impacts work revolves almost entirely around anticipating the effects of today's actions on a distant future's outcomes, either through planning and policy changes to reduce future impacts by adaptation or by mitigating emissions to avoid future impacts entirely. Thus prediction is central to the relevance of the IPCC and to the role of the modern climate science community in general.

The prediction of future outcomes is facilitated by a widespread use of computational models in the field. Scientists cannot perform controlled experiments on the Earth, so the effects of anything other than the realized emissions trajectory will never be observed. Instead, GCMs model the flow of mass and energy through the atmosphere to explore counterfactual scenarios and determine the potential effects of greenhouse gas forcing on climate. Coupled to these, integrated assessment models attempt to model both the human influence on climate through the energy system and climate change's influence on humans through the damage function. Both sets of models can be large, computationally intensive, and expensive to run and develop. They are also typically so complex that few, if any, developers have direct familiarity with all aspects of their model; entire careers can be devoted to understanding outputs and comparing diverse model results.

While developers and users of models tend to be aware of the simplifications and assumptions that prevent model outputs from reproducing the real world, personal and emotional investment in them often results in these inadequacies being overlooked (Lahsen 2005). A researcher might be simultaneously intensely aware of model shortcomings and intensely attached to model results as sound and meaningful scholarship, with different aspects of these conflicting attitudes manifesting in disparate contexts. For example, in some contexts modelers will insist that results are only projections (in other words, contingent on some set of inputs or boundary conditions and thus not necessarily comparable against real-world outcomes), while also insisting that results are relevant to policy planning as reasonable representations of future worlds. Similarly, modelers often refer to model runs as model experiments. Such terminology appropriates aspects of controlled experimentation, which lend legitimacy to so much of science, to the modeling enterprise and also indicates the general level of comfort with which climate scientists move between models and observations, the simulated and the real.

Despite the elaborateness and expense of general circulation of integrated assessment models, they do not attempt to achieve anything approaching the true complexity of the real world. Limitations on computing power mean the models still rely heavily on simplifying assumptions and extensive aggregation in ways that can seem absurd to the contextualized social sciences. For example, the Climate Framework for Uncertainty, Negotiation, and Distribution (FUND) model has one of the more sophisticated representations of impacts within the suite of integrated assessment models. Rather than reducing the impacts of climate change to a single damage function, FUND separately assesses impacts in twelve discrete sectors such as agriculture, tropical storm damage, vector-borne diseases, and energy consumption (Anthoff and Tol 2012). Nevertheless, tractability requires drastic simplification, even with this relatively sophisticated model. For instance, the model includes only sixteen regions around the globe, the only climate variables considered are regional temperature change and sea level rise, and impact sectors use very simple mathematical relationships between the outcome, climate variables, and per capita income to describe the impacts of climate change.

The contextualized social sciences are rightly wary of such heroic simplifications. The complex, interdependent, path-dependent, or emergent behavior in many of the systems studied means that reducing explanation to a suite of simple mathematical relationships to be used for long-term extrapolation is almost certainly unwise. A good cautionary example is that of long-term forecasts of U.S. energy use made in 1960–80. In a review Craig et al. (2002) found that modelers took the historical relationship between energy consumption and GDP as fundamental fact, failing to allow for improvements in energy efficiency as a result of the oil embargo and thus greatly overestimating consumption. Such examples are not hard to find, and they mean that social scientists are rarely, if ever, comfortable making predictions, particularly long-term predictions. But these are needed if the goal is, as it often is in much of climate science, long-term, quantitative prediction.

CHALLENGES FOR INTEGRATION

The preceding sections have discussed three elements of climate science that make integrating climate modeling work with the social sciences challenging: its spatial and temporal scale, its tendency to abstract changes related to climate change to changes stemming from other factors, and its attitude toward prediction. These challenges are not unrelated: they all stem from how climate change is defined as a statistical phenomenon, and all have similar implications not just for integration with the social sciences but for the broader acceptance of climate science as a legitimate basis for political action. This section discusses these commonalities and some potential avenues for overcoming challenges of integration and concludes with some thoughts on the attitude and response of climate scientists to insights originating in the literature of science and technology studies.

First, the three characteristics of climate science examined here are closely connected. For example, as discussed above, climate science is constrained to operate at large spatial and temporal scales because at smaller scales the signal of a trending climate is indistinguishable from natural variation due to chaotic weather patterns. But the importance of distinguishing this trend is a problem only if one's goal is to isolate the effects of a changing climate, either in historical weather records (such as in detection and attribution studies) or in future projections of changes. Similarly, the emphasis on prediction for climate policy–relevant impact assessment work is closely linked to the methodological importance of separating and quantifying impacts induced by climate change from others. The need to produce forecasts of likely future scenarios leads to a reliance on types of models that easily lend themselves to exploring the hypothetical and unobservable what-if baselines from which climate change impacts can be isolated and measured.

Taken together, these aspects of climate science aggregate into a set of methodological practices that can seem antithetical to the social sciences. Averaging over large spatial and time scales helps manage uncertainty owing to natural variability or small-scale effects that are not resolved in models and helps generalize findings across time and space. But an emphasis on statistical averages also strips away the social and environmental context in which climate change impacts become manifest. This becomes part of the unexplained variability or noise on observations, which are ignored for the purposes of impact projection. The social sciences, in contrast, tend to use thickly embedded descriptions of this contextual noise to manage uncertainty. Indeed, often this noise, which is just another name for the details of variation between individuals or communities, is of great interest to social scientists and is itself an object of inquiry. Particular observations are richly described, often with an emphasis on the elements that make these individuals or communities different from others. In this setting, uncertainty is managed by scaling down, not up, and through only limited and careful generalization of findings.

This trade-off between a thorough explanation of a particular phenomenon and the ability to generalize beyond those observations is well known in statistics, where it is known as the bias-variance trade-off. The fit of a statistical relationship is always improved by adding more explanatory variables, but the prediction error will increase if those additional variables are not structurally related to the outcome of interest. Analogizing to the various methodological traditions, we see that detailed contextual description (more explanatory variables) can increase understanding of a particular set of observations but possibly at the risk of less generalizability beyond these specific observations (higher prediction error). The climate sciences, in contrast, tend toward thinner descriptions that use fewer variables and that may have relatively low explanatory power but that can be more confidently applied elsewhere. There is no one solution to this trade-off, no universally optimal level of detail. Whether one prefers a highly detailed description and powerful explanation or better generalizability depends on the structure of the system under investigation and the type of prediction required.

A critical side effect of defining climate change as a statistical object, however, is that it does not admit the relevancy of individuals' or communities' own observations for knowledge production. In climate science, individual experiences gain legitimacy only when removed from context and combined with the experiences of others from different times and places to show a statistical climate change signal. So, for example, a community may experience a devastating drought or storm, but in order to tie such experience to climate change this event must be evaluated in terms of how it fits into the overall changing pattern of extreme events at spatial and temporal scales beyond the community's own experience. In such an environment, insights from the contextualized social sciences run the risk of being reduced to anecdote. They can add color and context to a conclusion but may struggle to persuade others, change conclusions, or substantially alter policy development. Overcoming this barrier to true integration will not be trivial since, as discussed above, it stems from the ways in which different disciplines manage uncertainty and, by extension, the definition of sound and persuasive fact.

Jasanoff (2010) has argued that the impersonal observations of climate science, because they operate at communal, political, spatial, and temporal scales larger than the embedded experience which legitimizes scientific knowledge, "have become decoupled from most modern systems of experience and understanding" (249). The fact-finding project of climate science has been separated from the meaning making of political institutions because it operates at scales larger than those at which the social processes that legitimize facts about the environment evolved. Jasanoff argues that this discordance between climate science and the political processes it seeks to inform "offers unique opportunities for disciplines that mainly concern themselves with the interpretive, sense-making capacities of human societies" (2010: 249).

Such meaning-making work is critically needed within the climate impacts and adaptation subfield. The situation today is one in which scientists broadly agree that unabated emissions will lead to a global rise in temperature of between 1.8 and 4.0 degrees this century, but the relevance of this fact to society remains both contested and deeply uncertain. Changing this will require a reembedding of scientific findings in the priorities and concerns of the communities they relate to, transforming what Latour (2004) has called "matters of fact" to "matters of concern." Such work represents, we believe, the most important and most exciting area of climate change research today. It offers the opportunity to develop new methodologies that can truly span the natural and social sciences, and it has the potential to produce new and timely insights into the relationship between humans and their environment.

Overcoming the barriers described above will be uncomfortable for those rooted in the traditional disciplines, but not impossible. It will require radically interdisciplinary research involving mixed methodologies and what Root and Schneider (1995) refer to, in the context of climate change impacts in ecology, as "strategic cyclical scaling"—the "continuous cycling between large- and small-scale studies, with each successive investigation building on previous insights obtained at all scales" (337). A hypothetical example of what such a mixed-methodology, cyclical-scaling study might look like when applied to the domain of climate change impacts on a social system is instructive (fig. 7.3). New and creative techniques can be developed but will require researchers from both the climate and the social sciences to work outside familiar traditional methodologies. For example, the climate sciences will likely have to revise the weight placed on quantitative, long-term prediction as something truly antithetical to a contextual, grounded understanding based on social science methodologies. Conversely, working with climate scientists may force social scientists to generalize findings to regional or national scales in ways they have not historically done.

Developing these new strategies for climate impacts research that integrates climate science and the social sciences will not be straightforward. But we believe these barriers to be primarily methodological rather than ideological. The climate science community has been faced with the demonstrable failure of the traditional view of science–policy interaction. The naïve model in which scientists develop a consensus on a problem and the best way of solving it, communicate that consensus to policy makers who then take action fails to explain the trajectory of the climate change issue. Direct political attacks on the integrity and professionalism of ordinary climate scientists have dramatically and urgently forced introspection within the discipline around the norms that govern the conduct and communication of research.

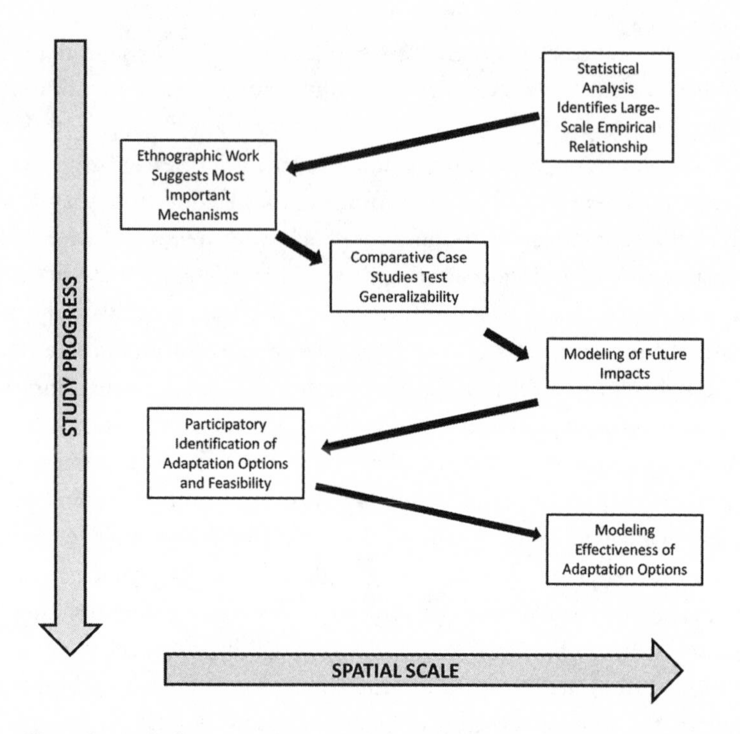

7.3: Hypothetical climate impacts. Frances C. Moore, Justin S. Mankin, and Austin Becker.

Moreover, the IPCC assessment process has grown to such an extent that by now a large minority of climate change researchers have participated in the process, with the most senior members typically taking part at a high level in at least two or three reports. Such experience cannot help but cultivate an appreciation of the subtleties of the science–politics interface. Such an appreciation, combined with frequent interactions with a skeptical public and opinion polls showing declining belief in climate change, is leading many to apprehend the limits of what knowledge founded only in the physical sciences can achieve (Leiserowitz et al. 2010). Natural scientists are therefore looking to the social sciences, both to explain the gap between the scientific and public perceptions of climate change and as the new frontier in knowledge and action regarding climate change impacts and adaptation. Indeed, some climatologists have moved entirely into the social sciences, and occasionally they are turning a critical eye, newly informed by science and technology studies (STS) theory, onto assessment processes they themselves were involved in (Hulme et al. 2009; Hulme and Dessai 2008).

Such questioning may represent a major opportunity for insights from STS and other social sciences to inform both the ways in which climate science operates and the ways in which it is presented to policy makers and the public. But taking full advantage of this opening will require going beyond academic critique to engagement by providing constructive and incremental suggestions for improvement. Speaking specifically to the case of global warming and the subversion of critical tools by organized skeptics to create an illusion of uncertainty, Latour asks whether it is possible "to transform the critical urge in the ethos of someone who adds reality to matters of fact and not subtract reality?" (2004: 232).

The climate scientist Stephen Schneider asked a substantively, if not rhetorically, similar question in reply to an STS article that pointed out the forces of social construction in both the identification of climate change as a global issue and the reliance on GCMs as the primary method of investigation (Demeritt 2001; Schneider 2001). Beginning with the phrase "I have long been swayed by the arguments of the Science and Technology Studies community that science is not the value-free, objective enterprise it often prides itself on being" (338), Schneider admonishes Demeritt principally for not producing a useful piece of work for "the climate scientists who are one of the prime target audiences of the effort" (341).

Similar questions are being asked as the IPCC, now a mature organization that has produced its Fifth Assessment Report, seeks to revise and improve its boundary-organizational role in order to better inform policy making in the future. What methodological, institutional, and communication changes would allow the organization to better achieve its mission of providing accurate, relevant scientific advice to policy makers and the public in the new age of global information networks but highly fragmented discourse?

These practical questions being asked today among senior climate scientists stem directly from the fact that, as Jasanoff (2010) has observed, climate science has as yet failed to integrate into social systems of knowledge legitimation. These questions represent just one of the openings for the type of constructive and creative engagement between the climate and social sciences that will be needed as climate change accelerates in the twenty-first century.

Climate change touches on almost all aspects of how we understand our world, and the need for cross-disciplinary work to understand it is frequently articulated. However, the definition of climate change as a statistical object makes meaningful integration with the social sciences challenging. People will experience climate change impacts at local scales, over human lifetimes, and embedded within a complex context of other ongoing social, political, economic, and technological change. But climate change is typically defined over much larger spatial and temporal scales and analyzed separately from other drivers of change. In this chapter we have argued that climate scientists' large scales of analysis arise at least partly from the need to manage uncertainty stemming from climate models and from the climate system itself. The tendency to abstract climate change impacts from their context and to value prediction into the future are disciplinary norms arising from an emphasis on quantitative modeling and the requirements of the political processes climate science seeks to inform.

As long as climate change impacts are understood only in a manner alien to the actual ways in which people relate to their environment, the threat posed by climate change is likely to be perceived as temporally and geographically distant, academic, and abstract. Changing this outlook requires an elaboration of the current thin and dry descriptions of climate change impacts into fuller and richer representations of how people engage with the environment and how a changing climate will impact society today and in the future. The social sciences are well placed to undertake this, but, because of the very different disciplinary norms discussed above, meaningful integration will likely involve a complex negotiation between research at the spatial and temporal scales of human experience and those larger scales at which climate science is most informative, as well as between specific and generalizable types of knowledge.

The value of research, its legitimacy, and its relevance are judged very differently, sometimes oppositely, in different fields. In the foregoing discussion, we have drawn on our experiences in proposing research and defending our approaches in the face of critiques from disparate epistemological traditions. We highlighted the unique challenges of applying knowledge generated at a large spatial and temporal scale to problems that today's decision makers can engage with on the local or regional level within the span of their own careers or lifetimes. Often, it is the type of system being studied and the kinds of data that can be obtained that constrain what type of work can be done. In this case, simply splitting the difference will likely produce something universally unsatisfactory. Instead, we believe strategic use of various methodological tools at a range of scales can produce rigorous and innovative scholarship that more effectively describes the effects of climate change.

Though work in climate change impacts and adaptation is providing the opening for innovative new methodologies that span disciplines and scales, the application of these techniques will likely stretch beyond climate change. The diffuse, multifaceted, and complex impacts that characterize climate change are likely to be features of other environmental problems of the twenty-first century. Developing the tools not only to quantify and analyze but also to make meaningful these types of impacts will be essential if we are to understand and tackle the environmental problems of the coming era.

NOTES

1. Jason Smeardon of Lamont-Doherty Earth Observatory to his graduate students.
2. There are local examples of climate change effects that are both obvious and clearly attributable (Orlove et al., this volume), particularly ice melt in the Arctic and the effect of sea level rise on low-lying island states, though even the latter is disputed (Webb and Kench 2010).
3. Similarly, projections for future sea level rise, already ranging widely about a global mean, are difficult to downscale to a level that is useful for decision making in any given location. A global mean of 0.6–1.9 meters could result in much greater or much lesser amounts of rise depending on local geomorphology and oceanographic processes. The implications for investment strategies at either end of this range can differ by millions or even billions of dollars.
4. http://www.cesm.ucar.edu/models/cesm1.0/timing_cesm1_0_5/.
5. The impacts section of AR5 is divided into sectoral chapters that seem principally defined by disciplinary or methodological differences. Thus the natural scientists using coupled models and end-point vulnerability dominate the "Natural and Managed Ecosystems" section, while social scientists using starting-point vulnerability perspectives are more prominent in the "Human Settlements, Industry and Infrastructure" and the "Human Health, Well-Being and Security" sections. Similar difficulties have been noted in the context of earlier assessment reports (Bjurström and Polk 2011).

REFERENCES CITED

Anthoff, D., and R. S. J. Tol. 2012. "The Climate Framework for Uncertainty, Negotiation and Distribution (FUND), Technical Description, Version 3.6." Accessed January 7, 2013. http://www.fund-model.org/versions.

Ayres, J., and S. Huq. 2009. "Supporting Adaptation Through Development: What Role for ODA?" *Development Policy Review* 27, no. 6: 675–92.

Barnett, T. P., D. W. Pierce, H. G. Hidalgo, C. Bonfils, B. D. Santer, T. Das, G. Bala, A. W. Wood, T. Nozawa, A. A. Mirin, D. R. Cayan, and M. D. Dettinger. 2008. "Human-Induced Changes in the Hydrology of the Western United States." *Science* 319, no. 5866: 1080–83.

Becker, A., S. Inoue, M. Fischer, and B. Schwegler. 2011. "Climate Change Impacts on International Seaports: Knowledge, Perceptions, and Planning Efforts Among Port Administrators." *Climatic Change* 110, nos. 1–2: 5–29.

Bjurström, A., and M. Polk. 2011. "Physical and Economic Bias in Climate Change Research: A Scientometric Study of IPCC Third Assessment Report." *Climatic Change* 108, nos. 1–2: 1–22.

Blaikie, P. M. 1985. *The Political Economy of Soil Erosion in Developing Countries.* Longman Development Studies, volume 21. New York: Longman.

Church, J. A., and N. J. White. 2006. "A 20th Century Acceleration in Global Sea-Level Rise." *Geophysical Research Letters* 33, no. 1: 94–97.

Craig, P., A. Gadgil, and J. Koomey. 2002. "What Can History Teach Us? A Retrospective Examination of Long-Term Energy Forecasts for the United States." *Annual Review of Energy and the Environment* 27, no. 1: 83–118.

Demeritt, D. 2001. "The Construction of Global Warming and the Politics of Science." *Annals of the Association of American Geographers* 91, no. 2: 307–37.

Deser, C., R. Knutti, S. Solomon, and A. S. Phillips. 2012. "Communication of the Role of Natural Variability in Future North American Climate." *Nature Climate Change* 2 (October): 775–80.

Diffenbaugh, N. S., M. A. White, G. V. Jones, and M. Ashfaq. 2011. "Climate Adaptation Wedges: A Case Study of Premium Wine in the Western United States." *Environmental Research Letters* 6, no. 2: 1–11. Accessed October 4, 2013. Doi: 10.1088/7148–9326/6/2/024024.

Edwards, P. N. 2010. *A Vast Machine: Computer Models, Climate Data, and the Politics of Global Warming.* Cambridge: MIT Press.

Engle, N. L. 2011. "Adaptive Capacity and Its Assessment." *Global Environmental Change* 21: 647–56.

Estrella, N., T. H. Sparks, and A. Menzel. 2007. "Trends and Temperature Response in the Phenology of Crops in Germany." *Global Change Biology* 13, no. 8: 1737–47.

Ferguson, J. 1994. *The Anti-Politics Machine: Development, Depoliticization and Bureaucratic Power in Lesotho.* Minneapolis: University of Minnesota Press.

Fogel, C. 2005. "Biotic Carbon Sequestration and the Kyoto Protocol: The Construction of Global Knowledge by the Intergovernmental Panel on Climate Change." *International Environmental Agreements* 5: 191–210.

Forsyth, T. 2003. *Critical Political Ecology: The Politics of Environmental Science.* London: Routledge.

Hansen, J., A. Lacis, D. Rind, G. Russell, P. Stone, I. Fung, R. Ruedy, and J. Lerner. 1984. "Climate Sensitivity: Analysis of Feedback Mechanisms." *Climate Processes and Climate Sensitivity* 5, no. 29: 130–63.

Hansen, J., L. Nazarenko, R. Ruedy, M. Sato, J. Willis, A. Del Genio, D. Koch, A. Lacis, K. Lo, S. Menon, T. Novakov, J. Perlwitz, G. Russell, G. A. Schmidt, and N. Tausnev. 2005. "Earth's Energy Imbalance: Confirmation and Implications." *Science* 308, no. 5727: 1431–35.

Hansen, J., G. Russell, A. Lacis, I. Fung, D. Rind, and P. Stone. 1985. "Climate Response Times: Dependence on Climate Sensitivity and Ocean Mixing." *Science* 229, no. 4716: 857–59.

Hawkins, E., and R. Sutton. 2009. "The Potential to Narrow Uncertainty in Regional Climate Predictions." *Bulletin of the American Meteorological Society* 90, no. 8: 1095–1107.

Hertel, T. W. 2010. "The Global Supply and Demand for Agricultural Land in 2050: A Perfect Storm in the Making?" *American Journal of Agricultural Economics* 93, no. 2: 259–75.

Hertel, T. W., M. B. Burke, and D. B. Lobell. 2010. "The Poverty Implications of Climate-Induced Crop Yield Changes by 2030." *Global Environmental Change* 20, no. 4: 577–85.

Hulme, M., and S. Dessai. 2008. "Negotiating Future Climates for Public Policy: A Critical Assessment of the Development of Climate Scenarios for the UK." *Environmental Science Policy* 11, no. 1: 54–70.

Hulme, M., S. Dessai, I. Lorenzoni, and D. R. Nelson. 2009. "Unstable Climates: Exploring the Statistical and Social Constructions of 'Normal' Climate." *Geoforum* 40, no. 2: 197–206.

IPCC. 2001. *Climate Change 2001: Impacts, Adaptation and Vulnerability.* New York: Cambridge University Press.

———. 2007a. *Climate Change 2007: The Physical Science Basis.* Edited by S. Solomon, D. Qin, M. Manning, Z. Chen, M. Marquis, K. B. Averyt, M. Tingor, and H. L. Miller. New York: Cambridge University Press.

———. 2007b. *Climate Change 2007: Impacts, Adaptation and Vulnerability.* Edited by M. Parry, O. F. Canziani, J. P. Paulutikoff, P. J. van der Linden, and C. Hanson. New York: Cambridge University Press.

———. 2012. *Managing the Risks of Extreme Events and Disasters to Advance Climate Change Adaptation: Special Report of the Intergovernmental Panel on Climate Change.* Edited by C. B. Field, V. Barros, T. F. Stocker, Q. Dahe, D. J. Dokken, K. Ebi, M. Mastrandrea, et al. New York: Cambridge University Press.

Jasanoff, S. 2010. "A New Climate for Society." *Theory, Culture and Society* 27, no. 2–3: 233–53.

Lahsen, M. 2005. "Seductive Simulations? Uncertainty Distribution Around Climate Models." *Social Studies of Science* 35, no. 6: 895–922.

Latour, B. 2004. "Why Has Critique Run Out of Steam? From Matters of Fact to Matters of Concern." *Critical Inquiry* 30, no. 2: 225–48.

Leiserowitz, A., E. W. Maibach, C. Roser-Renouf, N. Smith, and E. Dawson. 2010. "Climategate, Public Opinion, and the Loss of Trust." *SSRN Electronic Journal, 2010.* Accessed October 4, 2013. http://dx.doi.org/10.2139/ssrn.1633932.

Li, T. M. 2007. *The Will to Improve: Governmentality, Development, and the Practice of Politics.* Durham: Duke University Press.

Lorius, C., D. Raynaud, J. Jouzel, J. Hansen, and H. Le Treut. 1990. "The Ice-Core Record: Climate Sensitivity and Future Greenhouse Warming." *Nature* 347, no. 6289: 139–45.

Luers, A. L., D. B. Lobell, L. S. Sklar, C. L. Addams, and P. A. Matson. 2003. "A Method for Quantifying Vulnerability, Applied to the Agricultural System of the Yaqui Valley, Mexico." *Global Environmental Change* 13: 255–67.

Meyer, R. 2011. "The Public Values Failures of Climate Science in the US." *Minerva* 49, no. 1: 47–70.

Miller, C. 2001. "Hybrid Management: Boundary Organizations, Science Policy, and Environmental Governance in the Climate Regime." *Science, Technology and Human Values* 26, no. 4: 478–500.

Nelson, R., P. Kokic, S. Crimp, P. Martin, H. Meinke, S. M. Howden, P. De Voil, and U. Nidumolu. 2010. "The Vulnerability of Australian Rural Communities to Climate Variability and Change: Part II–Integrating Impacts with Adaptive Capacity." *Environmental Science and Policy* 13, no. 1: 18–27.

O'Brien, K., S. Eriksen, A. Schjolden, and L. Nygaard. 2004. *What's in a Word?: Conflicting Interpretations of Vulnerability in Climate Change Research.* Oslo: CICERO.

Oreskes, N., K. Shrader-Frechette, and K. Belitz. 1994. "Verification, Validation, and Confirmation of Numerical Models in the Earth Sciences." *Science* 263, no. 5147: 641–46.

Parmesan, C., and G. Yohe. 2003. "A Globally Coherent Fingerprint of Climate Change Impacts Across Natural Systems." *Nature* 421, no. 6918: 37–42.

Pierce, D. W., T. P. Barnett, H. G. Hidalgo, T. Das, C. Bonfils, B. D. Santer, G. Bala, M. D. Dettinger, D. R. Cayan, A. Mirin, A. W. Wood, and T. Nozawa. 2008. "Attribution of Declining Western U.S. Snowpack to Human Effects." *Journal of Climate* 21, no. 23: 6425.

Reidsma, P., F. Ewert, A. O. Lansink, and R. Leemans. 2009. "Vulnerability and Adaptation of European Farmers: A Multi-Level Analysis of Yield and Income Responses to Climate Variability." *Regional Environmental Change* 9: 25–40.

Root, T. L., and S. H. Schneider. 1995. "Ecology and Climate: Research Strategies and Implications." *Science* 269, no. 5222: 334–41.

Root, T. L., J. T. Price, K. R. Hall, S. H. Schneider, C. Rosenzweig, and J. A. Pounds. 2003. "Fingerprints of Global Warming on Wild Animals and Plants." *Nature* 421, no. 6918: 57–60.

Rosenzweig, C., and M. Parry. 1994. "Potential Impact of Climate Change on World Food Supply." *Nature* 367: 133–38.

Santer, B. D., K. E. Taylor, T. M. L. Wigley, T. C. Johns, P. D. Jones, D. J. Karoly., J. F. B. Mitchell, A. H. Oort, J. E. Penner, V. Ramaswamy, M. D. Schwarzkopf, R. J. Stouffer, and S. Tett. 1996. "A Search for Human Influence on the Thermal Structure of the Atmosphere." *Nature* 382: 39–46.

Schneider, S. H. 2001. "A Constructive Deconstruction of Deconstructionists: A Response to Demeritt." *Annals of the Association of American Geographers* 91, no. 2: 338–44.

Tett, S. F. B., P. A. Stott, M. R. Allen, W. J. Ingram, and J. F. B. Mitchell. 1999. "Causes of Twentieth-Century Temperature Change Near the Earth's Surface." *Nature* 399, no. 6736: 569–72.

Thorne, P. W., P. D. Jones, S. F. B. Tett, M. R. Allen, D. E. Parker, P. A. Stott, G. S. Jones, T. J. Osborn, and T. D. Davies. 2003. "Probable Causes of Late Twentieth Century Tropospheric Temperature Trends." *Climate Dynamics* 21, nos. 7–8: 573–91.

Webb, A. P., and P. S. Kench. 2010. "The Dynamic Response of Reef Islands to Sea-Level Rise: Evidence from Multi-Decadal Analysis of Island Change in the Central Pacific." *Global and Planetary Change* 72, no. 3: 234–46.

World Bank. 2010. "The Costs to Developing Countries of Adapting to Climate Change: New Methods and Estimates (Consultation Draft)." *The Global Report of the Economics of Adaptation to Climate Change Study.* Washington, D.C.: World Bank.

Zhang, X., F. W. Zwiers, G. C. Hegerl, F. H. Lambert, N. P. Gillett, S. Solomon, P. A. Stott, and T. Nozawa. 2007. "Detection of Human Influence on Twentieth-Century Precipitation Trends." *Nature* 448, no. 7152: 461–65.

Part Three **Imagining Climate Change**

Chapter 8 **Imagining Forest Futures and Climate Change: The Mexican State as Insurance Broker and Storyteller**

Andrew S. Mathews

The effect of climate change on contemporary imaginations is profound, including in the worlds of conservation and environmental anthropology. All problems, it seems, can be linked to and are indirectly caused by climate change. As we rush to carry out research on climate change and its impacts, it is worth thinking about other ways in which climate change has been imagined, and, especially, to think about how contemporary imaginations of climate change might make some kinds of things harder to think. In this chapter I compare two ways of connecting Mexican forests to climate change and national economies. My aim is to open up our imagination of climate change and reveal the power of imagination to remake what the state, climate, and forests are thought to be. The chapter is based on extensive ethnographic and historical research on Mexican forestry officials and indigenous forest communities (Mathews 2011) as well as on fifty-two recorded interviews on climate change and forest policies with forestry officials, scientists, and NGO representatives in Mexico City and Oaxaca, carried out in 2008 and 2009. Like McElwee (this volume), I find that Reduced Emissions through Degradation and Deforestation (REDD) policies are transformed by long-standing relationships between officials and rural people, and, like her, I use ethnographic research to compare official policies with what happens on the ground.

The first set of conceptual connections between climate change and forests, upon which I will touch more lightly, is that of regimes of conservation and industrial forestry that were established between 1920 and 1940, when forests were thought to prevent rainfall decline and droughts (Mathews 2009) and the science of knowing forests was one of measurement, calculation, and prediction. Although climate was understood to be affected by forests in general, certain species growing in particular forests were favored: these were pine trees growing in montane pine and pine–oak forests (Rzedowski 1978), which could produce industrial timber, understood as a volume and measured in board feet. Oak trees living in the same forests were little known, little measured, and largely unaccounted for and unpredicted. In the second moment, over the past ten years or so—the subject on which this essay focuses most attention—the Mexican state has increasingly begun to connect forests to climate change through speculative scenarios about landcover change, which underpin the emergence of REDD+ policies. In Mexico, REDD+ policy documents rely on classifications of forests according to their capacity to store or release carbon over time. Forests that contain long-lived, dense oak and other hardwood species may be more desirable than pine forests when calculated as tons of carbon stored across the landscape (Ordóñez et al. 2008), and in any case, oak presence—far from being uncounted—has become a valuable component of overall carbon stocks. Pines may become less desirable because they are less dense than oaks and, more important, because they are shorter lived.

Climate change can be imagined in many ways, whether in the former industrial forestry domain, in the domain of REDD+ scenarios, or in the domain of large climate change models. These different imaginings can fundamentally reconfigure how the state and its forms of action are imagined. By comparing how the state emerges in relation to forests and climate change at two separate moments, I draw attention not only to new forms of knowledge and action that arise but also to the kinds of things that become harder to think in each framing of state, forests, and climate. Oaks emerge as pines recede, the state comes to be understood not as a calculator and predictor but as an insurer of risk that elicits stories about the future. Crucially, state sovereignty and accountability take different forms when climate change is addressed through REDD+ models and policies. In the former scientific forestry formulation, the nation state was sovereign, responsible for regulating the nation's forests, establishing a credible regulatory framework for developing forest resources, and protecting citizens from the local and regional climatic impacts of deforestation, understood as floods and droughts (Mathews 2009, 2011). Although the founders of Mexican forestry in the 1930s were in communication with their international colleagues, they were acutely aware of their responsibilities to national publics and policy makers. In the new REDD+ formulation, credible knowledge about landcover change must be produced by independent scientists who are concerned with global carbon emissions. These scientists are in communication with their international colleagues and maintain a certain distance from the Mexican state. Further, the state must skillfully convene expert imaginations about future landcover changes, including through imagined disasters, in order to secure international funding (if this should ever arrive) for sustainable forest management programs to reduce carbon emissions.

In what follows, I will briefly describe how the state, climate, and forests were imagined in the 1930s and outline the regimes of forest management, prediction, and calculation that linked forest protection and timber production with the prevention of climate change. I then turn to the new forms of imagination and calculation by which forestry officials and landcover modelers seek to predict forest futures and make them a source of value for emerging REDD and existing Payments for Environmental Services (Pagos por Servicios Ambientales, or PSA) policies. Across this period, collective imaginations of the state are critical in stabilizing calculations about the future of forests. However, the role of the state has shifted from directly organizing such calculations to a more rhetorically modest position as the organizer and convener of scenarios about forest futures and, I argue, as the insurer of deforestation risk across the landscape.

CALCULATING REFORESTATION AND PERFORMING THE STATE

Forest protection has long been a state-mandated activity in Mexico, and numbers about reforestation and forest area have long been a feature of state environmental rhetoric. From the 1930s onward, under the minister of forests, fish, and game Miguel Angel de Quevedo, protection from the negative effects of deforestation on climate and the need to develop the national economy were the principal justifications for state forest protection (Mathews 2011). Since then, forests have been mapped and measured, becoming objects of calculation in order to plan local timber production, while national production statistics have been produced by collating timber transport authorization forms. Officials and presidents have recounted the environmental projects of the state through the recitation of statistics about numbers of trees planted, miles of roads built, and numbers of forest service offices established (Quevedo 1938). Statistical recitations of this kind have continued to be a feature of state environmental rhetoric, especially in the case of the mass reforestation campaigns that, as I discuss below, have taken on political significance.

Calculation and prediction have similarly been key features of the forest management plans that are mandated by the Mexican forest service. Here, forest management plans measure areas of territory, sample the numbers and sizes of (pine) trees, and predict the likely growth rate of forest for a period of five or ten years into the future (fig. 8.1). Such a plan works by measurement, calculation, and prediction. The state, whether as the producer of a forest management plan through the parastatal logging companies which dominated forests from the 1950s to the 1970s (FAPATUX 1977) or as the authorizer of the plan for the community forestry businesses that have dominated forests

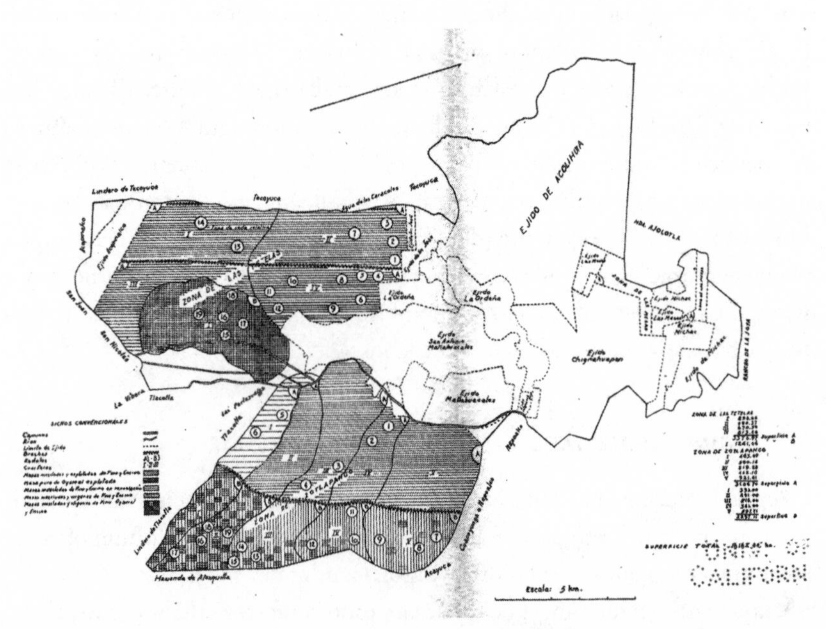

8.1: Forest management plan of Atlamaxa, Puebla, 1940. Treviño Saldaña.

since the mid-1980s (SARH and UCODEFO #6 1993; TIASA 1993), acts to ensure the successful calculation and prediction of forest growth. Other numbers, for example, those indicating tons of timber produced and area of forests across the nation as a whole, are likewise premised on counting and measurement (SEMARNAP 1997).

Forest management plans and statistics are helpful for managing forests, whether at the local scale or for entering national timber production statistics, but they also do other kinds of work. Such authoritative forms of knowledge not only declare what the world is but also do the work of performing the institutions that are responsible for acting in the name of this knowledge (Jasanoff 2004). Through the production of statistics of forest area planted, the Mexican state has performed itself as the kind of rational, ordered, and reasonably durable entity that gathers, processes, and acts upon knowledge of nature (Mathews 2008). Further, this is knowledge that is counted and calculated. Authoritative knowledge, then, is used to perform the state as calculator of present-day forests and, through its management plans and forecasts, to predict the future of forests over limited time horizons of a few years.

Practices of calculation do not take place in a void; they are made possible by a particular imagination of the state as a center of political and institutional order. The state is that menacing thing which is collectively imagined as somehow existing in time and space, while it is also justified and enacted by its knowledge performances.[1] Collective understandings of the state as being able to dominate national territory, transform indigenous people into modern citizens, and exercise occasional violence sustain the gathering and calculation of statistics about national deforestation and national timber production. Counting trees and forests and producing statistics about them both support and are supported by collective imaginations of a state which controls territory (as in the forest management plan above), guarantees property rights, and protects internal and external borders. A narrative of an overwhelmingly powerful, potentially violent state makes it possible also to imagine the state as a kind of rational entity, one which actually exists and which gathers statistical knowledge about nature in order to calculate and act on its predictions of probable futures. The state is performed through mundane rhetorical practices by forestry officials in the forest management plans they authorize and through the reforestation statistics they recite in meetings with indigenous forest landowners.

Mass reforestation campaigns in 2008–9 are a recent example of the performance of a calculating and predicting state. President Felipe Calderón committed the state to reforest hundreds of thousands of hectares and to plant millions of trees, with the goal of averting climate change. In the event, this mass reforestation program fell apart in the face of accusations of corruption and of dead and dying trees, energized by politically motivated counter audits which revealed the official reforestation campaign to be a failure (Mathews 2013). Calculation, it seemed, was not the best way of linking forests to climate change because there was widespread skepticism about the official statistics among critics, who pointed out that these were "doubtful figures" (Cevallos 2007) and who called for more definitive demonstrations of reforestation through satellite imagery. In fact, in parallel to the failed effort to produce national reforestation statistics through counting, measurement, and calculation, officials were engaged in bringing together independent scientists to model plausible and credible forest futures, which could be used in efforts to mitigate climate change. Such models of landcover change depend not upon calculation alone but upon a new understanding of the state as an entity which convenes scientists and technicians to produce scenarios about plausible future landcover change. These scenarios are a kind of empirical storytelling that positions the state as an insurance broker of deforestation risk.

REDD, REDD+, AND CLIMATE CHANGE

Over the past twenty years, an expanding host of actors has tried to link the biological capacity of growing forests to store carbon to efforts to slow the increase of atmospheric carbon dioxide. These policies, known as REDD and now REDD+ have emerged from the international United Nations Framework Convention on Climate Change (UNFCCC) treaty process (Forsyth and Sikor 2013). Acronyms such as REDD, REDD+, and AFOLU (Agriculture, Forestry, and Other Land Use) describe a variety of policies which fit under a broader REDD umbrella. What unites these policies is their shared goal of taking advantage of the ability of forests and agriculture to absorb carbon dioxide, given that land-use change contributes about 12 percent to global CO_2 emissions (Corbera and Schroeder 2011). As we shall see, REDD policies require the state to respond to new forms of knowledge and to elicit expert imaginations about plausible futures. These storytelling practices allow the state to become an insurer of deforestation risk.

In July 2008 a Mexican forestry official, Germán González Dávila, told me, "In the terrain [of REDD] Mexico could be one of the principal countries, not only to sell forest carbon [credits], but to offer the know-how of how to measure forest carbon. We are involved in that . . . we are trying to promote a megaproject for carbon cycling here in Mexico that could, if it got the necessary funds, generate information, practically in real time, from changes [in carbon storage] from twenty-some sample sites, in order to ground truth the changes that you are seeing by satellite" (recorded interview, Mexico City, July 22, 2008). Dávila was referring to the projects of a group of Mexican scientists who have pioneered the sale of carbon offsets by Mexican forest land owners in the state of Chiapas (de Jong et al. 2007; Ruiz-De-Oña-Plaza et al. 2011) and who are carrying out the research that will, they hope, make it possible to monitor Mexican forests by satellite. These scientists were partially funded by the British government, but their work has been largely embraced by Mexican policy makers and climate change negotiators, particularly at the Cancún climate conference in 2010. These scientists aim to measure a baseline level of carbon storage (Brown et al. 2007) so that possible carbon futures can be projected and compared, allowing for the sale of "additionality" should carbon offset markets take off. In the carbon world described by Dávila, real-time monitoring of forests produces scientific knowledge that is credible to international buyers, allowing money to flow from centers of finance to Mexican forests. Remote sensing of forests is attractive to policy makers because it is relatively cheap and also, as Dávila's words suggest, because seeing is a powerful metaphor for credible knowledge. These projects of vision are linked to the storytelling practices of scientists, through their production of scenarios about future states of the world. Crucially, the success or failure of REDD policies is measured not against actual forests but against counterfactuals, stories about how the world would have gone had the REDD policies not taken place.

For officials and scientists who wish to connect forests to markets, Mexican forests have to be mapped in the present so that their future states may be modeled as a trajectory of plausible futures, in order to sell the difference between a hypothetical scenario and the observed amount of carbon stored in forests and forest soils. Such projects might seem speculative and risky in the extreme, but they are anchored in practices of scenario building that are familiar to the financiers for whom the construction of a "business-as-usual scenario" is a routine skill, one taught in business schools. Critically, scenarios are not direct forecasts of probable futures; they are intended not to predict but to assist decision making under uncertainty, to describe not what is likely to happen but what could *plausibly* happen (Cooper 2010). Even as REDD policies are supported by a science of calculation, mapping, and measurement, they also rely on a science of storytelling, of looking back to measure several past moments of the world and then of turning to project this past into a range of credible futures. Modelers who try to predict how forests might increase or disappear in the future have to imagine what kinds of policies and events might affect forests, uncertain events whose probability is incalculable but which must be tamed by practices of principled doubt. Scenarios of landcover change are certified by international science and by the technical work of calculating coefficients of carbon storage and soil reflectance, but they also rely on correlations of landcover change with socioeconomic models, theories about society, social difference, and ethnicity. Modelers must judge the likelihood of socioeconomic changes if they are to come up with a scenario that is credible to their peers, to their employers, and to financial markets.[2]

Nightmares as Tradeable Value

Consider a model of a plausible future anchored in landcover modelers' and forestry officials' pessimistic visions, or even nightmares, as to what might happen in the world (fig. 8.2). This map offers a visual representation of the

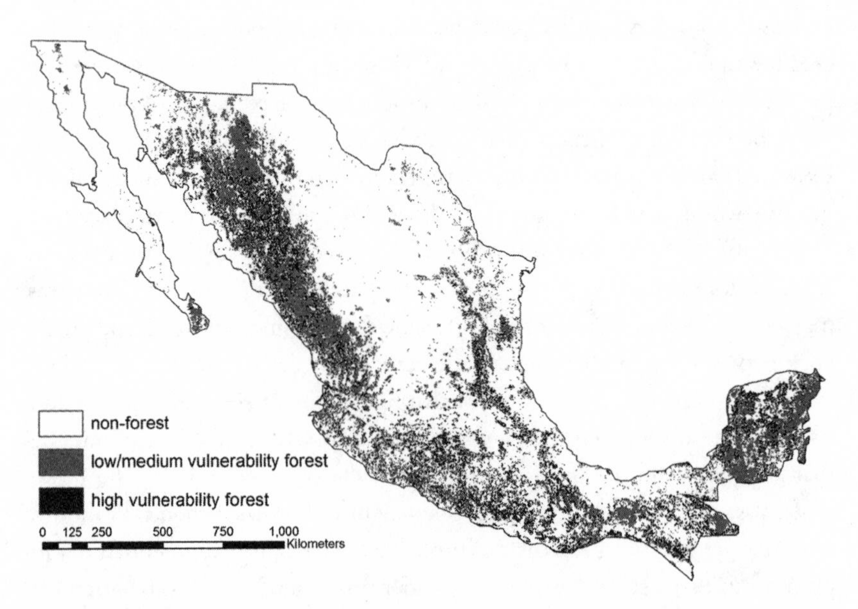

8.2: Vulnerable forests according to access and pressure. Redrawn by Lynn Shirley from the original by Ben de Jong.

probability of each pixel being deforested over a given period of time in the future, based on an assumed scenario about a plausible but incalculable future. This probability is based on a comparison of forest cover in 1990 and 2007 and a correlation of deforestation over this period with drivers (Grupo de Trabajo REDD México 2009). Access drivers include distance from roads, settlements, or developed areas. Pressure drivers include population density or other socioeconomic variables that index marginality or economic deprivation. (For details of this methodology by the same modelers, see Castillo-Santiago et al. [2007] and de Jong et al. [2005].) Modelers validate this retrospective analysis by predicting deforestation of a subset of their landcover data from 1990 and comparing predicted deforestation with what actually happened in 2007. The statistical success of this prediction helps imply the plausibility of a future-looking baseline scenario, but there is in fact, as the modelers are careful to warn (de Jong 2008), no way of establishing the credibility of a scenario from a retrospective calculation alone.

The probability of deforestation as posited in figure 8.2 is therefore linked to socio-environmental scenarios about the future (which is imagined to be similar to the past) in order to generate a calculation of the risk of deforestation. This calculated uncertainty is something that might be repackaged for markets as valuable risk, should REDD+ deliver on its promise of financial compensation for carbon storage across the landscape. These landcover change models could also change the value of particular tree species and forest and soil types, as the value of landscape in terms of carbon density could favor long-lived, dense tree species like oaks over less dense, rapidly growing pines (Ordóñez 2008). There is, however, some uncertainty on this point, as different coefficients for converting vegetation type to carbon density are in use (de Jong et al. 2010).

As depicted in figure 8.3, forest at risk of deforestation is correlated with forest that has high "social importance." The original image explains that the factors which determine social importance are rural employment, population density, marginality, primary sector activity, and the presence of indigenous people. Nightmares about future environmental degradation caused by poor indigenous people are linked to baseline scenarios of likely future deforestation that may, through the linkage of social factors to calculated deforestation risk, underpin business plans and budgets. The presence of indigenous people has become a social risk factor, one that can be linked to the risk of deforestation and represented, as here, in a map. In outlines of national REDD policies and in policy presentations, maps of endangered forest, maps

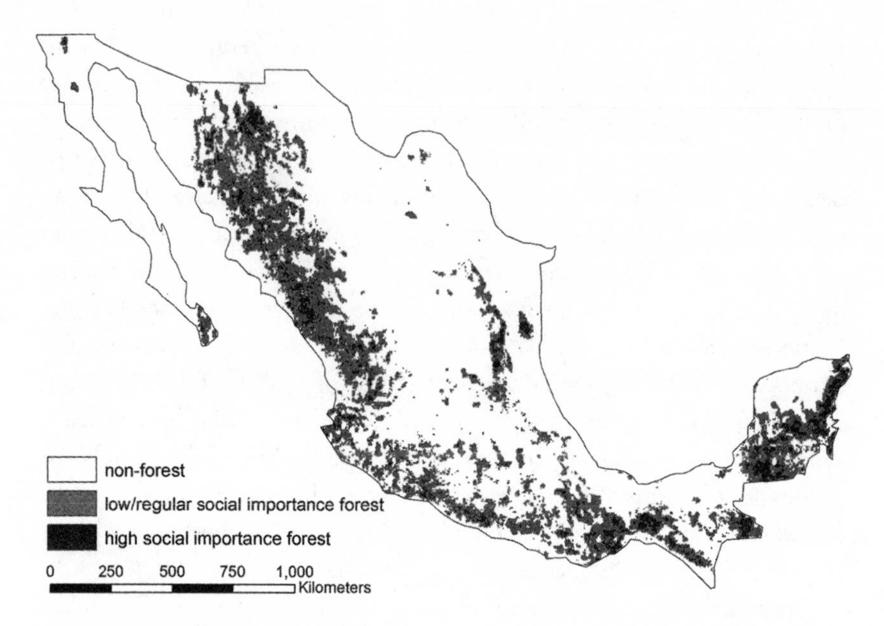

8.3: Forest in relation to social factors. Redrawn by Lynn Shirley from the original by Ben de Jong.

of socio-environmental risk, and maps of the location of indigenous people rapidly follow one another as the presenters advocate investment in capacity building and technical training to implement REDD policies in Mexico (Gutiérrez et al. 2008; Coordinación General de Producción y Productividad Gerencia de Servicios Ambientales del Bosque 2009). Curiously, once indigenous people are treated as an environmental risk, they become a potential source of value. How does this surprising result emerge?

Indigenismo is an official indicator that records the percentage of people who are willing to tell a census taker that they speak an indigenous language or who otherwise self-identify as indigenous. In Mexico, government officials take care to note how indigeneity and another official index, marginality (*marginación*), are highly correlated. Marginal places are typically far from roads, and the populations who inhabit them have low levels of literacy and little access to running water, electricity, etc. Marginality makes poor indigenous people ideal subjects for pessimistic scenarios about the quantity of carbon stored on future landscape. These people live in places where deforestation has occurred in the past and (according to one version of the future) where it is likely to continue, perhaps owing to growing populations or to the supposedly destructive impacts of firewood cutting and agropastoral burning on forests. The more pessimistic the initial measurement of forest carbon and the more pessimistic the vision of the future embedded in the scenario, the higher the potential financial value this scenario holds. When you succeed in persuading indigenous people not to cut down forests or in preventing them from doing so or you entice them to abandon their agricultural fields, you have made a new, measurable carbon world. The difference between the world as it is thus measured and the world you imagined in your business-as-usual scenario (or baseline) is the amount of carbon "stored." This is what you hope to sell to international carbon markets in averted emissions from forests, if these markets should ever be built. Perversely, pessimistic scenarios about indigenous people fly in the face of considerable evidence that indigenous forest communities in Mexico are *less* likely to deforest the environment than other kinds of landowners (Bray et al. 2004).

A nightmare of indigenous people destroying the environment becomes more valuable as it becomes more nightmarish, with an added caveat: the international money will not arrive unless you act to make the nightmare go away. You have to change the livelihood practices of indigenous people in order to make them stop farming or cutting down forest. In effect, indigenous people, regarded sympathetically by the modelers and officials who wish to channel REDD finance toward them, are targeted by a scenario that requires indigenous people to change what they do.

It would be easy to imagine that REDD plans are a kind of authoritative administrative transparency that will hurt poor people, forcing them to abandon their productive practices as some parts of the landscape become unavailable to food production. Certainly, this is a real concern. However, like the regimes of calculation and counting embedded in forest management plans and mass reforestation, REDD regimes have shown signs of being vulnerable to the imaginations and political organizing of audiences who imagine both the state, which administers REDD programs, and the technological knowledge of the landcover change that underpins REDD programs.

The reasons for the shift from REDD to REDD+ policies, which have increasingly come to the fore since the Cancún climate conference in 2010 (Martin and Margaret 2011), are particularly telling in this regard. REDD, in theory, was supposed to compensate *averted deforestation*, through the practices of scenario planning I have outlined above. REDD+, on the other hand, will reward a host of practices, including improved forest management, reduced forest degradation, and landscape-level carbon storage. In REDD policies the state coordinates the production of technical information on landcover change, but money is supposed to flow to the property owner identified by the cadastral map, with relatively little state intervention. Under REDD+ policies, the state reemerges as the organizer and supporter of sustainable agricultural and forest management practices, including conservation, across the landscape. Organizing and convening landscape-level monitoring and measurement of carbon stocks take place under both REDD and REDD+, but the value of forest carbon is less directly captured by individual landowners in REDD+. In this model, the state acts as a source of subsidies and as an insurer of deforestation risk across the landscape because it is at a regional or national level that total tons of increased carbon storage will be demonstrated to financial backers, should there ever be a market in emissions reductions. The rationale behind the shift from REDD to REDD+ in Mexico illustrates the striking vulnerability of these state-making projects to international events. The shift to REDD+ was negotiated at Cancún, partially in response to the demands of forest-rich countries but also because of the vulnerability of state forestry projects to national political pressures and popular imaginations.

One reason for the shift from REDD to REDD+ has been the logistical difficulty of mapping individual property ownership and of establishing a tight connection between an individual property, an owner, the carbon stored in that owner's forests, and payments for averted deforestation. One environmental NGO in the state of Oaxaca, Servicios Ambientales de Oaxaca (SAO), has taken what might be called a pure REDD approach in what is known as the voluntary carbon market, in which private individuals or companies can offset their carbon emissions by buying certificates of averted deforestation from forest-owning indigenous communities. Technicians from SAO map the boundaries of indigenous forest communities, maintain a map database that links each map polygon to tons of carbon stored, and train community forestry technicians to measure trees and monitor their growth. This strategy reduces but does not eliminate the presence of the state, which still certifies the carbon credits issued by SAO. Organizations which try to implement such pure REDD strategies must confront, in a particularly acute form, the problem of leakage: that is, the deforestation of nearby land not included in the REDD program. They must also face the possibility of the sudden loss of carbon into the atmosphere through a forest fire or disease outbreak. The technicians deal with these possibilities in three ways: by using below-ground carbon as an unquantified buffer to ensure overall additional carbon storage, by maintaining a buffer of forest in each community that can be brought into the program if an area within the program is cleared, and by maintaining a buffer at the level of multiple forest communities. In other words, a baseline scenario of carbon storage is present and is used by forestry technicians, but the absorption of additional carbon overall is insured by SAO, which is able to make and break conceptual and material relationships between areas of forests and the scenario. At this point, SAO, with state support, is, in effect, constructing a deforestation risk pool so as to insure against the risk incurred by the buyer of the emission reduction certificate. Because this transaction takes place in the voluntary carbon offset market, the nature of the risk is mainly reputational. A buyer would presumably not wish to be seen to have purchased a carbon offset from a fraudulent or failed program, but they are not actually incurring the loss of property that they own. In any case, although conceptually there is a neat link between a payment and a landowner, in practice SAO, like other sellers of forest carbon credits, stands ready to substitute alternative areas of forest if an area within the program should be cleared for agriculture or destroyed by wildfire.

Rivals and critics of SAO, both academic researchers and officials, told me that this approach was unfeasible owing to the high costs of mapping and maintaining a property and payments database, arguing that this model would be uneconomic at the large scale of Mexico's forests. However, the limitations of a pure environmental services market approach are also clear in the Mexican government's PSA program, run by the national forestry commission, Comisión Nacional Forestal (CONAFOR), and funded by the World Bank and the Mexican government. The PSA program carries out the same kind of insurance of deforestation risk that has been performed by SAO. Because the PSA does not have enough money to enroll all of the applicants' area of forest into the program, it enrolls only a portion of a given landowner's forest. If a landowner in fact clears an area of forest, state-level forestry officials substitute the cleared area for intact forest that had not been enrolled in the program. This is an entirely sensible decision by officials, who wish to encourage rural people to protect forest, not punish them for infractions. However, this forbearance is probably also inspired by officials' awareness of the limited legitimacy and stability of the PSA program in the wake of a scandal over reforestation that nearly destroyed a different branch of CONAFOR in 2008–9 (Mathews 2013). The effect of this practice of "swapping" forest into and out of the program is, however, to produce precisely the kind of leakage that REDD programs are supposed to avoid. In this case, leakage is produced not by landowners but by officials who link new areas of forest with the PSA program. In effect, additional offsite deforestation, which could be called leakage in REDD terms, has been made invisible.[3]

This invisibility is not a defect but is intentionally produced by government officials, who act to insure against the risk of a particular area of forest being deforested, and in so doing also insure against the reputational risk to the forestry commission and its World Bank backers. There is an enormous scholarly literature on insurance, including the ways in which modeling practices produce new forms of insurable and hence valuable risk (Johnson 2011). However, what I am focusing on here are the mundane bureaucratic practices of officials who pool the risk of deforestation across the landscape in order to insure against the deforestation of any particular area of forest. These insurance practices for now sustain only the *reputation* of carbon offset purchasers, but such practices would also insure financial value if REDD+ policies were ever implemented so as to produce financial markets in carbon credits. Although no true REDD+ program as yet exists in Mexico, it is likely that these kinds of insurance practices would also be necessary to support regional REDD+ programs if they were ever to be put into practice. In such a case, carbon stored at a regional or landscape level would be more tightly tied to a model of deforestation risk. This model would underpin a deforestation risk pool, which would be guaranteed by the regional- or national-scale forestry institutions that insure deforestation risk across the landscape over which they hold authority.

Mexican environmental organizations have been skeptical of REDD approaches, partly because of their high administrative costs, given that much of rural Mexico lacks the kind of detailed cadastral property map on which REDD approaches are premised. There is, however, an entirely different set of reasons that has led radical environmental organizations to push back hard at the state environmental services program. Since I began work on REDD in 2008, environmental NGOs supportive of rural communities have heavily criticized the functioning of the existing PSA program, forcing forestry officials to modify not only the program but also Mexico's official negotiating position on REDD+. A very active community forestry and environmental NGO sector has pressured the Mexican state to make sure that climate change policy channels funds into forest communities that either conserve their forests or manage their forests carefully through sustainable logging practices. These communities are the charismatic clients of state-sponsored forestry in Mexico. Indigenous forest communities, among whom I have carried out much research, are highly effective forest managers, the poster children for international researchers like me and for eliciting funding from sympathetic donors like the World Bank. Through political pressure, occasional protests, and pacific occupation of government offices indigenous forest communities and their NGO allies have greatly modified the PSA program (McAfee and Shapiro 2010), ensuring that forest communities *who are already* protecting their forests will be compensated if REDD+ ever gets off the ground.

Although compensating presently occurring conservation and forest management practices flies in the face of the additionality that REDD programs are supposed to ensure, forestry officials have rapidly moved to REDD+ formulations, through which they can more easily channel resources toward their long-term allies and interlocutors. This is in part because of the political weakness of forestry officials, who need the credibility and glamor that the NGO and community forestry sector can provide. This was brought home to me in the spring of 2011 in a striking way at a conference on indigenous community forestry in Oaxaca. I had given a presentation on REDD approaches and explained how such a pure environmental markets approach would make it difficult to compensate forest communities for the work they had already been doing in protecting their forests because there would be no additional carbon stored. Almost immediately, a nervous official announced to his audience of community leaders and environmental activists that REDD+ would compensate the indigenous forest community sector. The state, which had rhetorically disappeared in REDD policy models, becomes more visible in REDD+ approaches, where the state acts as an insurer of deforestation risk across the landscape (because it is at a regional level that additionality is to be measured), while it allows officials to maintain relationships with the NGOs and forest communities who support or criticize the forest service. Although there is as yet no national REDD+ program in Mexico, the practices of officials in the existing PSA program suggest that a healthy buffer will allow officials to turn a blind eye to forest clearance or rule breaking by rural people. Dangerous forest futures are tamed through the state's ability to switch an area of forest that has been cleared over here (a pixel on a satellite image) for another area of uncut forest over there (a different pixel) or, better still, for the total tons of forest stored across the landscape at a regional or national level.

COMPARING CALCULATION AND SCENARIO BUILDING

A comparison of two regimes of climate–forest connections, first in the mid-twentieth century and then in the present, makes visible how each regime increases or decreases the value of certain tree species and forest types depending on the coefficients through which trees are converted into numbers. Formerly, the numbers of interest to the state were timber volume in cubic meters, making pine trees valuable and causing oak trees to be almost entirely ignored because they did not produce saw timber. In the regimes of REDD+, things are less clear because the number of concern is tons of carbon per hectare, and this number is arrived at through individual species and soil-specific coefficients (de Jong et al. 2010). The exact interaction of forests with soils makes calculations uncertain, but the oak species in mixed pine oak and oak forest types (Rzedowski 1978) become more valuable, as they contribute to total carbon stored across the landscape (Ordóñez et al. 2008). Rapidly growing species such as pines, which are logged and rapidly contribute their carbon back to the atmosphere, may become less valuable and less important. All forms of lower-value landscape transformation and agropastoral work become problematic, both because they produce greenhouse gas emissions and because they are a source of uncertainty for modelers who seek to calculate carbon fluxes. Given the likelihood that the officials responsible for running a REDD+ program in Mexico will be acutely aware of reputational risk to their institution, they will probably maintain an ample buffer of carbon storage and, in keeping with long-term official policies, avoid the counting and controlling of agropastoral burning or rural firewood use (Mathews 2011).

The forest service is the branch of the Mexican state that came into being in relation to mass reforestation and forestry during the 1920s. At this time, the nation-state located uncertainty and danger in particular places, such as at the borders of the nation-state and in internal ethnic enemies, that is, the indigenous people who were targeted by state-sponsored projects of education, modernization, and cultural *mestizaje.* These programs tried to make indigenous people into modern Mexicans, while forestry regulations tried to protect forests from their agricultural and pastoral practices. The state tried to prevent climate change by requiring foresters and landowners to measure, count, and calculate the extent and future growth of forests over the five- or ten-year cycles of forest management plans. Individual forest management plans and transport documents were collated into national statistics of timber production and became one of the ways through which the state performed itself before national audiences.

Notably, the state that makes such calculation possible is not something that itself can be calculated. Rather, it is a collectively imagined entity, one which collects the energies of the nation, controls national territory, and travels through time, more or less solid and coherent, into the future. Such a state makes itself accountable to the public under the conditions of calculation. It asks to be measured by the quality of its statistical predictions about the area of forest, of roads and schools built, and of numbers of trees planted. Over the past century in Mexico, the state has tried to calculate and predict the trajectory of future forests in order to develop forests for the national economy or to protect them for conservation purposes. Crucially, collective imaginations of the state itself, as a fearful and knowledgeable actor, were necessary to make calculation and prediction about timber production in forest management plans even provisionally credible. Calculation was sustained by imaginations about politics, authority, and the nature of the state.

Practices of scenario planning which underpin REDD policies demand a fundamentally different orientation toward the future on the part of officials and those who might criticize them. First, REDD policies require not that the state directly collect national timber production or deforestation statistics but that it convene the technical knowledge of landcover modelers. Such modelers are, however, kept at arm's length, so that the knowledge they produce can attract international financial backing and scientific credibility. If protecting forests from indigenous people in the 1930s required a negative image of their relationship to a nation-state that directly gathered and calculated knowledge of forests, channeling finance into forests in the present requires that scenarios about indigenous people be built by modelers who are at some distance from the officials who orchestrate their research. Here, the supposedly degrading practices of indigenous people become a possible source of value, but only if they are linked to credible socio-ecological scenarios. In this new formulation, the state is imagined as a more or less humble convener of the technical knowledge of scientists and scenario builders.

A scenario is a kind of creative imagination of a story about the future that provides the basis for calculation and prediction. Thus a scenario is a principled engagement with future uncertainty and deals with events whose probability simply cannot be calculated (Cooper 2010). In making scenario planning a central feature of state response to bioterror preparedness (Masco 2014), climate change, or other inherently unpredictable occurrences, the state asks to be held accountable not to predictability but to plausibility and to the quality of its practices of convening the imaginations of experts into scenario building. Should a disaster happen, officials can say, "We developed an appropriate process for producing the best kinds of scenarios possible." So far, most scenarios of landcover change have consulted the imaginations of landcover modelers themselves (e.g., Brown et al. 2007) and of forestry officials (Pelletier et al. 2011), but not, to my knowledge, of rural or indigenous people. If REDD+ models were ever to be translated into actual efforts to control the behaviors of indigenous people, such efforts would be profoundly unjust. However, as we have seen, REDD+ projects are largely speculative, and Mexican officials at least are likely to be careful to avoid enforcing restrictive regulations among rural people.

In the past, numerous anthropologists and like-minded critics have sought to unmask the pretensions to objectivity and certainty that are produced in great abundance by modern states through such practices as cost-benefit analysis and risk analysis (Lohmann 2009). Scenario making about landcover change might seem similarly scandalous, containing as it does the desires of powerful actors who try to remake the present in the name of their imaginations of the future. I argue, on the contrary, that there is nothing inherently scandalous about calculations that are sustained by imaginations of plausible socio-environmental futures. Imaginative scenarios about plausible futures are found not only in REDD modeling practices, but also in climate change models and across a broad domain of the natural sciences that deal with complex, multicausal problems. What is of interest here, then, is the fact that REDD modeling so clearly reveals the necessarily imaginative and political content of *all* scenario building, including the scenario-building practices of climate modelers, who with their paraphernalia of supercomputers and sophisticated calculations might seem to be very far removed from the relatively mundane practices of landcover change modelers in Mexico. Imagination is properly constitutive of our common world. Hence my concern is not to reveal this as a scandal but to compare the diverse locations of imagination and calculation under scientific forestry and REDD modeling. What this comparison shows is that practices of calculation and prediction always require collective imaginations of what authoritative institutions are and can do (Jasanoff 2004). Climate modeling, like REDD modeling and like the forest management plans of industrial forestry in mid-twentieth century Mexico, call for collective imaginations of the political institution that can act in the name of the knowledge which they produce.

Let me turn now to the question with which I began. What is made harder to think and notice in this new practice of connecting forests with climate? It seems to me that REDD policies make political authority emerge in relation to the production of value, which is generated in turn by averting scenarios of environmental degradation. By comparison, under state-sponsored industrial forestry, resources were channeled from the periphery into the center, and information about how to manage forests went in the other direction. In REDD+ policies, should these ever come to be, the state would insure the funder or buyer of carbon credits against the risk of a particular area of forest suffering from fire, deforestation, etc. and would distribute subsidies to people who live and work in forests and fields. Through its control over territory, the state would mediate flows of funds into forest protection and agricultural policies and would summon the scientific knowledge about landscape produced by credible third parties. A way of thinking that focuses on the flow of funds *from* the state to the margins systematically obscures flows of information and resources from the periphery *to* the state.[4]

Comparing former regimes of calculation and industrial forestry with those of insurance and scenario making in the present loosens, a little, the power of climate change on our collective imaginations. It was only when I began to compare changing flows of resources and information between forests and climate change in the regimes of industrial forestry that began in the early twentieth century and in REDD policies that have come to the fore in the past ten years that a third possibility came to mind. In focusing so intensely on where clearing or growing forests might cause the absorption or emission of carbon dioxide, our present-day way of imagining climate–forest connections obscures the place that the vast majority of carbon is being emitted *from*. These are the urban centers and power plants where much carbon is entering the atmosphere in the first place. If we stretch our imaginations still further, we could imagine long-ago forests that were buried and turned into coal and that are now brought into the atmosphere through coal mining and burning. Perhaps a comparison of the area of carboniferous forest implied by fossil fuel burning would make REDD projects aimed at relatively modest areas of present-day forests seem less sensible, less obviously attractive.

NOTES

1. There is a long tradition of considering how calculation is made possible by practices of commensuration within the social studies of finance (Callon 1998). As Jane Guyer points out, calculation can be made possible by skillful performances, which link

different incommensurable scales of value (Guyer 2004). Here, I suggest that the menacing presence of the state is an additional element in establishing the possibility of the calculations that enter forest management plans.

2. It is also possible that the potential buyers of carbon credits are simply uninterested in whether actual carbon is removed from the atmosphere, so that the crucial business-as-usual scenario is no more than a demonstration of due diligence (Lohmann 2009).
3. In actual fact, because most of the rural communities enrolled in the PSA program have declining involvement in agriculture and did not intend to clear their forests anyway, there is little deforestation and relatively little leakage (Alix-Garcia et al. 2010).
4. This is an intriguing inversion of the pattern described by Dove and Kammen (2001) in Indonesia.

REFERENCES CITED

Alix-Garcia, J. M., E. N. Shapiro, and K. R. E. Sims. 2010. "Forest Conservation and Slippage: Evidence from Mexico's National Payments for Ecosystem Services Program." *Staff Paper Series*, University of Wisconsin, Agricultural and Applied Economics.

Bray, D. B., E. A. Ellis, N. Armijo-Canto, and C. T. Beck. 2004. "The Institutional Drivers of Sustainable Landscapes: A Case Study of the 'Mayan Zone' in Quintana Roo, Mexico." *Land Use Policy* 21: 333–46.

Brown, S., M. Hall, K. Andrasko, F. Ruiz, W. Marzoli, G. Guerrero, O. Masera, A. Dushku, B. DeJong, and J. Cornell. 2007. "Baselines for Land-Use Change in the Tropics: Application to Avoided Deforestation Projects." *Mitigation and Adaptation Strategies for Global Change* 12, no. 6: 1001–26.

Callon, M. 1998. "An Essay on Framing and Overflowing." In *The Laws of the Markets*, ed. M. Callon, 244–68. Oxford: Oxford University Press.

Castillo-Santiago, M., A. Hellier, R. Tipper, and J. DeJogl. 2007. "Carbon Emissions from Land-Use Change: An Analysis of Causal Factors in Chiapas, Mexico." *Mitigation and Adaptation Strategies for Global Change* 12, no. 6: 1213–35.

Cevallos, D. 2007. "Ambiente Mexico: Reforestación Cosecha Detractores." *Inter Press Service*. July 10. Accessed October 28, 2013. http://www.ipsnoticias.net/2007/07/ambiente-mexico-reforestacion-cosecha-detractores.

Cooper, M. 2010. "Turbulent Worlds: Financial Markets and Environmental Crisis." *Theory, Culture and Society* 27, no. 23: 167–90.

Coordinación General de Producción y Productividad Gerencia de Servicios Ambientales del Bosque. 2009. "Estrategia Nacional de REDD–en Preparación-Reducción de Emisiones Derivadas de la Deforestación y Degradación Forestal." Comisión Nacional Forestal, Mexico.

Corbera, E., and H. Schroeder. 2011. "Governing and Implementing REDD+." *Environmental Science and Policy* 14, no. 2: 89–99.

De Jong, B. 2008. "Reducing Emissions from Deforestation and Degradation: Example of a Regional Reference Scenario and Advances of a National Approach in Mexico."

UNFCCC Workshop on Methodological Issues Relating to Reducing Emissions from Deforestation and Forest Degradation in Developing Countries. Zapopan: CONAFOR.

De Jong, B., C. Anaya, O. Masera, M. Olguin, F. Paz, J. Etchevers, R. D. Martinez, G. Guerrero, and C. Balbontin. 2010. "Greenhouse Gas Emissions Between 1993 and 2002 from Land-Use Change and Forestry in Mexico." *Forest Ecology and Management* 260, no. 10: 1689–1701.

De Jong, B., E. Bazán, and S. Montalvo. 2007. "Application of the 'Climafor' Baseline to Determine Leakage: The Case of Scolel Té." *Mitigation and Adaptation Strategies for Global Change* 12, no. 6: 1153–68.

De Jong, B., H. J. A. Hellier, M. A. Castillo-Santiago, and R. Tipper. 2005. "Application of the 'Climafor' Approach to Estimate Baseline Carbon Emissions of a Forest Conservation Project in the Selva Lacandona, Chiapas, Mexico." *Mitigation and Adaptation Strategies for Global Change* 10, no. 2: 265–78.

Dove, M. R., and D. M. Kammen. 2001. "Vernacular Models of Development: An Analysis of Indonesia Under the 'New Order.'" *World Development* 29, no. 4: 619–39.

FAPATUX. 1977. "Estudio Dasonómico de las Comunidades de San Juan Bautista Atepec y Miguel Aloapam." Oaxaca: 178.

Forsyth, T., and T. Sikor. 2013. "Forests, Development and the Globalisation of Justice." *Geographical Journal* 179, no. 2: 114–21.

Grupo de Trabajo REDD México. 2009. "Escenario de Referencia (Componente7-VersiónBorrador Ecosur)" 16. Mexico City: Colegio de México.

Gutiérrez, I. L. 2008. "Policy Approaches and Positive Incentives on Issues Relating to REDD in Developing Countries; and the Role of Conservation, Sustainable Management of Forests and Enhancement of Forest Carbon Stocks in Developing Countries." *UNFCCC AWG-LCA/Policy Approaches and Positive Incentives Relating to REDD in Developing Countries.* Zapopan: CONAFOR.

Gutiérrez, I. L., J. A. A. de la Rosa, et al. 2008. "Advances of Mexico in Preparing for REDD." *UNFCCC Workshop on Methodological Issues Relating to Reducing Emissions from Deforestation and Forest Degradation in Developing Countries.* Zapopan: CONAFOR.

Guyer, J. I. 2004. *Marginal Gains: Monetary Transactions in Atlantic Africa.* Chicago: University of Chicago Press.

Jasanoff, S. 2004. *States of Knowledge: The Co-Production of Knowledge and Social Order.* London: Routledge.

Johnson, L. 2011. "Climate Change and the Risk Industry: The Multiplication of Fear and Value." In *Global Political Ecology,* ed. R. Peet, M. Watts, and P. Robbins, 185–202. London: Routledge.

Lohmann, L. 2009. "Toward a Different Debate in Environmental Accounting: The Cases of Carbon and Cost-Benefit." *Accounting, Organization and Society* 34, nos. 3–4: 499–534.

McAfee, K., and E. N. Shapiro. 2010. "Payments for Ecosystem Services in Mexico: Nature, Neoliberalism, Social Movements, and the State." *Annals of the Association of American Geographers* 100, no. 3: 579–99.

Martin, H., and S. Margaret. 2011. "Monitoring, Reporting and Verification for National REDD+ Programmes: Two Proposals." *Environmental Research Letters* 6, no. 1: 014002.

Masco, J. 2014. "Pre-Empting Biosecurity: Futures, Threats, Fantasies." In *Bioinsecurity and Vulnerability*, ed. N. Chen and L. A. Sharp, 5–24. Santa Fe: School of Advanced Research Press.

Mathews, A. S. 2008. "Statemaking, Knowledge and Ignorance: Translation and Concealment in Mexican Forestry Institutions." *American Anthropologist* 110, no. 4: 484–94.

———. 2009. "Unlikely Alliances: Encounters Between State Science, Nature Spirits, and Indigenous Industrial Forestry in Mexico, 1926–2008." *Current Anthropology* 50, no. 1: 75–101.

———. 2011. *Instituting Nature: Authority, Expertise and Power in Mexican Forests.* Cambridge: MIT Press.

———. 2013. "Scandals, Audits and Fictions: Linking Climate Change to Mexican Forests." *Social Studies of Science* 40, no. 1: 82–108.

Ordóñez, J. A. B., B. H. J. de Jong, F. Garcia-Oliva, F. L. Avina, J. V. Perez, G. Guerrero, R. Martinez, and O. Masera. 2008. "Carbon Content in Vegetation, Litter, and Soil Under 10 Different Land-Use and Land-Cover Classes in the Central Highlands of Michoacan, Mexico." *Forest Ecology and Management* 255, no. 7: 2074–84.

Pelletier, J., N. Ramankutty, and C. Potvin. 2011. "Diagnosing the Uncertainty and Detectability of Emission Reductions for REDD+ Under Current Capabilities: An Example for Panama." *Environmental Research Letters* 6, no. 2: 024005.

Quevedo, M. A. d. 1938. "Las Labores del Departamento de 1935 a 1938." *Boletín del Departamento Forestal y de Caza y Pesca* Year 3: 39–79.

Ruiz-De-Oña-Plaza, C., L. Soto-Pinto, S. Paladino, F. Morales, and E. Esquivel. 2011. "Constructing Public Policy in a Participatory Manner: From Local Carbon Sequestration Projects to Network Governance in Chiapas, Mexico." In *Carbon Sequestration Potential of Agroforestry Systems*, ed. B. M. Kumar and P. K. R. Nair, 247–62. Netherlands: Springer.

Rzedowski, J. 1978. *Vegetación de México.* Mexico City: Editorial Limusa.

SARH and UCODEFO #6. 1993. "Programa de Manejo Forestal para la Comunidad de Pueblos Mancomunados de Lachatao, Amatlán, Yavesía y Anexos, Mismos Municipios Distrito de Ixtlán de Juárez, Oax. Oaxaca, México." Secretaría de Agricultura y Recursos Hidráulicos.

SEMARNAP (1997). "Anuario estadístico de la producción forestal." Mexico City: Secretaría de Medio Ambiente y Recursos Naturales y de Pesca.

TIASA. 1993. *Programa de Manejo Forestal, Ixtlán de Juárez, Oax., 1993–2000.* Texcoco.

Treviño Saldaña, C. 1937. "Proyecto de Ordenación de los Bosques de Atlamaxac, Puebla." *Boletín del Departamento Forestal y de Caza y Pesca* 8: 177–253.

Chapter 9 Digging Deeper into the Why: Cultural Dimensions of Climate Change Skepticism Among Scientists

Myanna Lahsen

In fieldwork I initiated in 1994 I set out to do an ethnographic study of climate modelers. Basing myself at the U.S. National Center for Atmospheric Research (NCAR) in Boulder, Colorado, I wanted to study the atmospheric scientists who produce and use the complex computerized general circulation models (GCMs). The GCMs simulate variations in atmospheric variables over time and under different scenarios and inform climate policy because they project possible climate futures owing to human emissions of greenhouse gases.

In common anthropological fashion, I could not stay put. I was compelled to follow how the scientific knowledge flowed in both scientific and nonscientific contexts (Martin 1998), thereby rendering my research multisited. I traveled around the United States to interview scientists in institutions similarly involved with climate science, including a subset of the "contrarian" American scientists. In this chapter, I relate key findings from this research, focusing on subcultural tensions that structure climate science politics in the United States. I connect the obviously politicized and staunchly critical climate skeptics—whom I call contrarians—with more moderate, low-profile skeptical views I encountered within the scientific mainstream.

The focus of my fieldwork expanded in synch with my growing fascination with the broader scientific and political contexts in which the climate models are produced and discussed. In the lead-up to the international adoption of the Kyoto Protocol under the United Nations a large number of newspaper articles and industry-funded campaigns emerged opposing climate policy (Lahsen 2005b; McCright and Dunlap 2000; McCright and Dunlap 2003) and attacking the the United Nations Intergovernmental Panel on Climate Change (IPCC). The Contract with America campaign in 1995, led by Rep. Newt Gingrich, secured the Republicans' majority in the 104th Congress, bolstering the anti-environmental movement and weakening environmental legislation (Kraft 2000; Lowry 2000). In this context, the House Subcommittee on Energy and Environment held a series of hearings to investigate publicly funded environmental research, including activities supporting the IPCC; the hearings challenged scientists who were concerned about climate and prominently displayed contrarian scientists and viewpoints. The Republican subcommittee chairman, Rep. Dana Rohrbacher, publicly derided climate science as "scientific nonsense" and global warming as unproven "liberal claptrap" (Lawler 1995).

In my typology, mainstream scientists work in official scientific institutions (mainly accredited universities and federal research laboratories), obtain research funding from government agencies, and publish primarily, if not exclusively, in scientific, peer-reviewed journals (table 9.1). Mainstream skeptics share these characteristics. By contrast to contrarian scientists, mainstream skeptics are moderate in their questioning of the science underpinning worries about human-induced climate change. They tend to believe that global climate has warmed and that human action may be among the causes, but they question aspects of the evidence and are critical of what they perceive to be exaggerations of the threat of climate change and of its scientific certainty. Unlike contrarians, they tend not to challenge the evidence of other environmental problems like ozone depletion and acid rain, and they do not have material and discursive ties to the vested interests and conservative think tanks that propel the anti-environmental movement. They also lack contrarians' strong, explicit aversion to government regulation. While application of these definitions is admittedly difficult and ambiguous in some instances, the typology serves to characterize broad divergences among the subgroups.

Table 9.1. Differences in tendencies between mainstream climate concerned scientists, mainstream skeptics, and contrarians (adapted from Lahsen 2013: 3)

	Mainstream climate concerned scientists	Mainstream skeptics	Contrarians
Question evidence of anthropogenic climate change	Only parts, at the most, not the theory as a whole	Moderately so, yes, including the theory as a whole	Strongly, categorically
Skeptical of many other issues of environmental concern (planetary limits, pollution, etc.)	No	No	Yes
Work in official scientific institutions (accredited universities, federal research laboratories)	Yes	Yes	In some cases/ partly
Primary venue for writings and publications	Peer-reviewed scientific journals	Peer-reviewed scientific journals	Non-peer-reviewed outlets (newspapers, reports, blogs, etc.)
Material and discursive ties to conservative factions	No	No/rarely	Yes, always
Political values	Liberal	Liberal	Conservative

Credit: From *American Behavioral Scientist*

The scientists I studied often appeared in the media, especially contrarian scientists, who invariably supported—indeed, were central instigators of—controversies related to climate science that erupted regularly in the national media. They were key actors in backlash productions such as the television documentary *The Greenhouse Conspiracy* (1990) and its sequel of 2007, titled *The Great Global Warming Swindle.* Both productions aired on public broadcasting television channels in Britain and elsewhere, and they continue to enjoy a global audience via YouTube and other Internet sites. They have been viewed by millions of people, despite strong criticisms that they misused and fabricated data, relied on out-of-date research, employed misleading arguments, and misrepresented positions of the IPCC and individual scientists. Staunchly critical of the notion of anthropogenic climate change and presenting climate modelers as being willfully ignorant of empirical data, the documentaries posited climate scientists as driven by ulterior motives. The narrator of *The Greenhouse Conspiracy* indicts the media as well as an alleged overdramatization by scientists, highlighting especially Stephen Schneider and Tom Wigley, both former NCAR scientists, and suggesting that their beliefs reflect hugely expensive but error-prone climate models whose outputs are unsupported by observations.[1] A segment of the documentary ends with an earnest, puzzled experimental meteorologist from MIT, Reginald Newell, saying, "I don't know why models are taken seriously."

To promote their agenda, powerful backlash actors have frequently adopted deceptive strategies to create fictitious appearances of broad grassroots and scientific support (Lahsen 2005b). Nevertheless, they have at times managed to also draw on a small but not insignificant number of PhD scientists, including a subset of mainstream environmental scientists. Newell's appearance in *The Greenhouse Conspiracy* is a case in point, as are a number of compilations of scientists' names on open letters or declarations expressing skepticism of climate science and associated policy initiatives.[2]

Criticisms of GCMs pervade the articulations of contrarian scientists, whose conservative inclinations are obvious and well documented. However, in line with the backlash documentaries and signatory lists, I also heard criticisms of climate models and of the preoccupation with climate change among lower-profile mainstream scientists.

Like the contrarians, the more moderate mainstream critics, whom I refer to as mainstream skeptics, represented a variety of scientific fields and lines of practice. However, as my fieldwork progressed, I came to note patterns in who they were. I perceived common generational dimensions to the conflicting views regarding climate change and climate models and noted that this generational aspect, along with a few other features, connected these skeptics to the high-profile contrarian scientists. I especially observed an older generation of climatologists who resented and regretted how times had changed meteorology, and I found scientists in related environmental fields echoing their complaints. I came to understand that their criticisms, regardless of their truth-value from a scientific perspective, reflected a competition for status and research funds. I detected a relative marginalization and alienation in them in relation to changes in their field, changes associated with the rise of concern about new, far-reaching but uncertain and future-set environmental problems. A paradigm shift within meteorology privileged theory and simulation techniques, and the field grew more obviously politicized and policy oriented. I argue that the positions of climate contrarians and subgroups of skeptically inclined mainstream scientists reflect reactions to these changes, reactions that are cultural in nature and thus transcend individual characteristics.

A subset of the high-profile climate contrarians have passed away since my fieldwork in the 1990s and 2000s, and some new actors have joined the climate backlash. However, contemporary climate science politics in the United States confirms the continued relevance of insights gained and actors and dynamics identified during this fieldwork. As late as 2008, two of the actors discussed in detail below—Frederick Seitz and Fred Singer—participated in another backlash campaign, a scientific report issued under the name "Nongovernmental International Panel on Climate Change" (NIPCC). Repeating common backlash tenets, the report debunked climate models and criticized environmental policy. Indicative of the international influence of the contrarians, I learned about the NIPCC report through an article published by a Brazilian contrarian scientist, José Carlos de Azevedo, in the Brazilian newspaper *Folha de São Paulo*, as I was living in Brazil by that time. Azevedo explicitly cited Seitz as a scientific authority, mentioning Seitz's positions as past president of the American Academy of Sciences and of the American Physical Society to lend force to his own contrarian views (Azevedo 2008).

THE STUDY OF CULTURE IN GLOBAL ENVIRONMENTAL CHANGE RESEARCH

There have been few analyses of the cultural contradictions generated by climate change and other broad-scale human transformations of the global environment. Through acts of abstraction and avoidance, climate science continues to commonly appear to be independent of the culturally laden specificities of human experience in its production and reception (Demeritt 2011; Jasanoff 2010; Lahsen 2010; Proctor 1998). Reflecting and perpetuating this lacuna, social science in the IPCC tends to be limited to economic studies and to human responses to changes in natural environments (Demeritt 2009). Values hardly enter the analysis (Duraiappah 2010), and sociological analyses of science are entirely absent. For their part, social scientists have produced relatively little research on specifically *cultural* dimensions shaping climate science and policy gridlock (Lahsen 2010), including the role of lived experience, or "habitus" (Bourdieu 1991), despite suggestions that effective policy responses may hinge on such analyses (Hulme 2008; Hulme 2009).[3]

Analyses of the dynamics related to global environmental change tend to confirm the cultural theory framework initially developed by Mary Douglas and Aaron Wildavsky (1984). Focusing on individuals, they confirm the structuring role of core values identified in the cultural theory framework, namely, the extent to which individuals hold individualistic, hierarchical, communitarian, and egalitarian values. In line with other studies (for example, Douglas et al. 1998; Leiserowitz et al. 2010; Pendergraft 1998; Silverman 2010; Thompson and Rayner 1998), Dan Kahan (2010) recently concluded that differences in these core values explain divergences in climate risk perceptions more completely than do differences in gender, race, income, educational level, political ideology, personality type, or any other individual characteristic. He writes, "People with individualistic values, who prize personal initiative, and those with hierarchical values, who respect authority, tend to dismiss evidence of environmental risks, because the widespread acceptance of such evidence would lead to restrictions on commerce and industry, activities they admire. By contrast people who subscribe to more egalitarian and communitarian values are suspicious of commerce and industry, which they see as sources of unjust disparity. They are thus more inclined to believe that such activities pose unacceptable risks and should be restricted" (Kahan 2010: 296).

Anthony Leiserowitz et al. (2013) similarly note that these core values reflect individuals' values, wishes, and preferences and that they influence the information to which people pay attention, how they evaluate data, and the conclusions they draw. They conclude that egalitarians, by contrast to individualists, are predisposed to perceive climate change as a serious risk and to support a variety of policies to address it.

The cultural theory framework appears applicable to climate science politics. Analyses of the dozen or so high-profile contrarian American scientists have identified their personal individualistic and conservative values (Lahsen 1998; Lahsen 2008; Oreskes and Conway 2010). However, I contend that the positions of the various scientists and subgroups involved also reflect shared (cultural) dispositions. The deeper sociocultural dimensions are revealed only by paying attention to the subgroups' relationships with each other and in associated meaning-making drawing on shared particularities of experiences and historical memory.

Any given individual's behavior is overdetermined by a mix of idiosyncratic and shared, for example cultural, factors (Mouffe 1988), and climate skepticism is similarly overdetermined (see Lahsen 2008). Accordingly (and because of space limitations), my analysis does not pretend to be exhaustive; it focuses specifically on the role of scientific values and experiences related to GCMs in structuring contrarian and mainstream skeptics' attitudes and engagements related to climate change.

HABITUS AND TRANSFORMATIONS IN SCIENCE

In my research I identified three subgroups of mainstream scientists who question climate change. The first two are kinds of research meteorologists: the dynamicists, who by definition are theoretical and strong in physics, and the more empirical climatologists, especially older scientists who at some point in their lives worked in weather prediction. The third group is formed by weather forecasters.

Some history of meteorology will help orient those who are unfamiliar with these subgroups. Meteorology historically subsumed three lines of practice: weather forecasters, climatologists, and dynamicists. Members of the latter two groups do research on climate dynamics. Climatologists make use of theory, but they are empirically oriented. Dynamicists use data but concentrate primarily on theoretical dimensions of atmospheric dynamics. I specifically reserve the label *dynamicist* for research meteorologists who have not adopted modeling techniques. Theory was weak in meteorology until the development of numerical methods secured a new role for dynamicists in forecasting, which before that point had been produced by means of the empirical "synoptic" methods. Synoptic methods involved forecasting of atmospheric dynamics in time and space from the disparate points from which meteorological data were collected. Synopticians, those who specialized in this method, drew on their empirical knowledge of how the atmosphere moves in varying locations and conditions to connect the dots. That is, doing what is now done by models, they would infer the movements of air and other meteorological variables between the points from which data were available. By contrast to modeling, the synoptic method was not physics-based, however. It draws on intuitive, experience-based understanding of how the atmosphere works, albeit informed where possible by the laws of physics.

The numerical methods ended synopticians' role as forecasters, and their use of weather maps for research was likewise gradually replaced by numerical methods. The numerical methods were deemed superior because they were based on quantitative methods, unlike the synoptic methods, which henceforth came to be referred to as subjective methods. A new branch of meteorology was born with the numerical methods: climate modeling. Its emergence opened a greater space for physics, although some theoreticians continue to work with theory rather than with the numerical models. Some climatologists—specifically, the subset of empirical meteorologists—who had worked by applying synoptic methods in the analysis and forecasting of weather, have since begun using models to perform tasks previously done by means of weather mapping (which does not make them into modelers, a term I reserve for those doing GCM climate projections).

Particularities of their habitus render especially older members of the three traditional subgroups of meteorology relatively more inclined to be critical of models. Their lived experiences, scientific training, and practice give them reasons to doubt pronouncements made on the basis of GCM output. Older members of these subgroups have often experienced a demotion or, at least, a reduction in their access to research funds relative to those using numerical climate models. Aside from expertise-derived insight, alienation in response to these changes and to the associated broader transformations in science and society can thus underpin skepticism espoused by members of these subgroups. This is a key point of connection between them and contrarian scientists, whose climate-related views are similarly inflected with alienation and discontent (Lahsen 2008).

New criteria have emerged for evaluating what constitutes good and worthwhile science. Knowledge production used to be based in discipline, involved a clear distinction between fundamental and applied science, and placed the greatest value on basic research, understood as a necessary precursor of applied science and engineering. Pure theory, physics, and mathematics served as the ideals in this paradigm, in which searches for first principles and unitary theory of the world are primary goals. Good science was equated with that which allows prediction and control with high precision and single-variable measurement and manipulation.

But a new sociopolitical context and set of methods have given rise to a competing value system in which disciplinary boundaries and common distinctions between fundamental and applied science are less important.[4] Knowledge in this newer paradigm is often explicitly linked to policy goals—indeed, research is seen as having value in great part for consideration of its practical, societal, and policy-related impacts. The search for first principles is a less primary or exclusive driver of research than interest in understanding concrete systems and processes and addressing issues of societal concern (Gibbons et al. 1994). Federal agencies' funding criteria have shifted increasingly in favor of the new paradigm since the end of the Cold War. In the environmental sciences, this has translated into higher funding levels for so-called socially relevant science, which has privileged research on climate change and for which GCMs are the essential tool.

In a clear articulation of the new paradigm, Schneider, an outspoken proponent of policy-driven, interdisciplinary climate research, including modeling, expressed in an interview with me his lack of interest in "an elegant solution." He unequivocally defended the search for answers to pressing social problems by any means necessary, including imprecise science, finding a preference for detail and greater precision unethical given the high environmental and societal stakes in reaching a greater understanding of the potential consequences of current trends in greenhouse gas emissions. Schneider also expressed his opinion that while "knowledge for its own sake is fine, too . . . that should only get a smaller fraction of the pie."

By contrast, the contrarians and mainstream skeptics whom I discuss below show strong marks of the old paradigm in their scientific orientations and values. They have often been negatively impacted by the rise of the numerical models and the environmental and scientific focus on human-induced climate change. Indeed, Schneider often features as a target in their criticisms precisely because he, more than many others, symbolized the changes in science and society that affected them negatively or which they otherwise view with regret.

CONTRARIANS

The Physicist Trio

A physicist elite that joined under the conservative Marshall Institute, a think tank based in Washington, D.C., was led for over a decade by the now-deceased Seitz, William Nierenberg, and Robert Jastrow, scientists who were extraordinarily active and influential in the climate backlash (Lahsen 2008; Oreskes and Conway 2010). The physicist trio's habitus were formed before the contemporary wave of environmentalism, at a time when communism was perceived as a dominant threat and nuclear physicists enjoyed the highest prestige in the scientific hierarchy (Lahsen 2008).

At least at the discursive level, this physicist elite mobilized traditional scientific values in their criticisms of the environmental sciences and, in particular, of climate modeling. Defenders of basic science and having a value framework squarely rooted in the postwar decades and the Cold War mentality, they were at odds with the growing environmental movement and the associated values that gained force during the 1970s.[5] These transformations challenged their social status and influence as well as the types of science they most valued. By contrast, climate modeling and other lines of science that these physicists defined as inferior enjoyed promotion. Their backlash engagements expressed their dismay and resistance to these changes in science and society and a defense of their understandings of science, of modernity, and of themselves as a physicist elite (Lahsen 2008). Adding to the fire, these new environmental concerns were advanced by a coalition of actors who also rejected the nuclear technology they and their physicist mentors had helped develop and that they defined as being crucial to national security and economic prosperity (Lahsen 2008; Oreskes and Conway 2010).

During a field trip I made to the East Coast in 1995, I interviewed Seitz in his large office at Rockefeller University in New York City, where he was president emeritus. He quickly embarked on negative portrayals of modelers as opportunistic technicians:

SEITZ: It started with the same group that in the seventies was warning us of a new ice age. They got a lot of publicity and then they seized on global warming—and got a lot of publicity. Also, they became well funded for computer research, for computer modeling. . . . I associate the original outburst—if you want to call it that—with a man called Schneider who is on the Stanford faculty. He's a brilliant computer operator.

Seitz's reference to Schneider as being brilliant was only a way of softening what in fact was an attack: the label "computer operator" was an insult in Seitz's value hierarchy, which celebrates basic science and elegant solutions in the form of equations. Seitz described modelers as diverging from "standard procedure" in science, as power- and publicity-seeking "true believers . . . who believe the world is coming to an end" and who enjoy dictating others' lifestyle.

LAHSEN: You see scientists doing this [dictating how people should live]?

SEITZ: Scientists joined by others. Then you have a group of people who are enjoying good funding for ingenious experiments with computers. They call them experiments, but they are not tied necessarily to observations out there, in the real world. And of course they would like to see their funding continue, so they find it convenient. . . . I come out of the traditional attitude towards science that ultimately you have to use observations as your base, then combine it with speculation and theory, then see where you come out. To date, there is no significant evidence that we're in impending danger.

Seitz expressed his support of the science policy that marked the decades following the Second World War. By contrast to the present, in his rendition, funding priorities were then made on the basis of needs of science without much regard to "political issues." While this may have been Seitz's perception, historical analyses show that science funding during postwar decades in the United States were similarly structured by politics, more specifically by foreign policy objectives related to the Cold War (Forman 1987; Kevles 1990). Indeed, Seitz and his physicist colleagues were centrally involved in related policy processes as advisors. The scenario of funding the most gifted scientists and granting them total freedom in their scientific pursuits was more obvious in the 1930s and identified with Andrew Carnegie, in particular (Naomi Oreskes, personal communication).

Seitz described the scientists with whom he associated himself: scientists who value and practice traditional scientific techniques, by contrast to climate modelers:

SEITZ: Please understand that I am part of an extended, international group [of people] who has the same feeling—that we have got to use the methods of good science on these issues.

LAHSEN: Do you want to expand on that? Who are these people?

SEITZ: Well, we come out of the traditional base of science, which is, you know, the techniques of science that have been built up with great care and sharpness over centuries. The ultimate resolve of issues is experiment.

LAHSEN: When you say you want to use good science, what is it that proponents of global warming do that you don't agree with? Where do they diverge from what you call traditional science?

SEITZ: They are doing computer runs, which often disagree violently with the observations. And they then rush out to the press, the media, claiming this or that, without proper scientific justification.

LAHSEN: Okay so again, the science that is being done right now, you say that it is not good science because it is not based enough on observations, right?

SEITZ: That's right.

LAHSEN: So, inherently about models, you would say that it is not a very scientific method?
SEITZ: Yes.

Seitz placed climate models at the center of his science-focused attacks and, as evidenced in the quote above, especially criticized Schneider as a scientist who represents a kind of science and values that conflict with his own. In his biography Seitz similarly reproduced and expressed dismay at Schneider's public statement that to engender desired societal responses in the face of environmental threats scientists "need to offer up scary scenarios" and "make simplified, dramatic statements" that downplay doubts, a statement Seitz associated with "extremist prejudice" and attitudes "not uncommon at the present" (Seitz 1994: 382). Seitz lamented political correctness on university campuses, in particular, the move away from Western classics toward multiculturalism. Asked how these criticisms related to our conversation, he explicitly linked both to the rise of computer modeling and to new paradigms in science and society: "They say it is time for a whole new outlook. Science came out of Western civilization, which [they say] was bad—what we did to the Indians, and so forth. And in their view it is time to have a whole new approach. And maybe the computers are the solution!" Criticisms of GCMs are thus especially strong among contrarians, where they mesh variously with neoconservative notions of the New Class and political correctness.

Demoted Research Meteorologists and Contrarianism

A theoretical research meteorologist specialized in radiation theory, Hugh W. Ellsaesser was retired when I interviewed him on two occasions in the latter half of the 1990s. Early in his career, around the time of the Second World War, he was trained in synoptic methods and operational weather forecasting. In public writings Ellsaesser has, like Seitz, evoked the generational aspect, portraying both models and modelers as young and unreliable: "The optimum attributes for developing [climate models] are burning ambition and an uncluttered mind—which helps explain why most such models have been developed by graduate students. I do not believe that most of us would agree that such people are in general the ones who best understand how the atmosphere as a whole works" (Ellsaesser 1989: 71). In an interview Ellsaesser expressed the less polemical criticism that the models have become black boxes. He noted that GCMs can be freely downloaded and used yet come unaccompanied by documentation about assumptions, simplifications, and biases they contain, undermining their meaningful use in advancing scientific knowledge.

Ellsaesser's arguments show convergences with conservative groups with which he has associated, including the followers of Lyndon LaRouche, a controversial, American political activist and founder of the right-wing-linked LaRouche movement associated with extreme views and inclinations toward conspiracy theories. He shares with LaRouche a fearful emphasis on new international structures such as the United Nations and sees climate change as a plot by which the United Nations propagates an ideology of political correctness and aims to take over the United States with a world government. Ellsaesser is characteristic of the contrarians in the skepticism he shows with regard to many other environmental issues, including lead poisoning, air pollution, and the use of DDT. He understands environmentalism as being caused by a new generation's inability to correctly perceive problems, much less solve them (Ellsaesser 1992).

The factor of their politically conservative individual values apart, contrarians such as Seitz (not an atmospheric physicist) and Ellsaesser (a theoretical meteorologist with past training in forecasting methods) share central elements of their criticisms and of their scientific profile with mainstream subgroups of skeptically inclined scientists. A common denominator of Ellsaesser and a subset of more mainstream skeptically inclined scientists, namely, an older generation of climatologists, is the fact of having been trained in weather forecasting early in their careers, at a time when forecasting was accomplished via synoptic methods rather than numerical models. Ellsaesser identifies himself as a weather officer and lists his twenty-one years with the U.S. Air Force Weather Service among his credentials.

What distinguishes Ellsaesser from his skeptically inclined fellow research meteorologists is his extreme political conservatism and his engagements with the backlash, which isolated him in his position at the Lawrence Livermore National Laboratory, a federal entity. When I interviewed him in his laboratory in November 1996, I found him occupying a single office in a trailer full of otherwise empty offices. Two trailers immediately next to his were packed with mainstream environmental scientists. Ellsaesser's physical location in the trailer by himself indicated the complete breakdown in his communication with the scientists in the other trailers and with the mainstream scientific community as a whole; all existent communication was antagonistic and took the form of politicized commentaries transmitted through the mass media. My conversations at Lawrence Livermore revealed that past attempts at communication had only impressed on both sides their inability to agree with each other.

NONCONTRARIAN, SKEPTICALLY INCLINED EMPIRICAL METEOROLOGISTS

Mainstream Empirical Meteorologists

Representing a new hybrid form of scientific inquiry, climate modeling gradually assumed a central role in the atmospheric sciences, to the point of sidelining empiricists and theoreticians. At U.S. research institutions like NCAR, the tasks of all scientists were increasingly made subservient to the climate modeling enterprise. Moreover, despite the need for all three groups—empiricists, theoreticians, and modelers—to produce good models, in practice (as revealed through my fieldwork), empiricists and theoreticians feel that their knowledge and expertise are not always integrated into that enterprise. Modelers tended to be young scientists and to have more mathematical know-how than knowledge of the actual atmosphere (Harper 2003), a fact indicative of a new emphasis on theoretical knowledge and modeling in doctoral programs in atmospheric science. The space for more pure observationalists and for synoptic methods has shrunk considerably; today, synoptic methods are taught mainly to forecasters at the master's rather than the doctoral level.

One strand of the criticisms of models that I encountered during my fieldwork expressed some scientists' association of them with a feared "loss of thinking" in science owing to the overreliance on simulation technology, a fear identified as well in historical research on nonenvironmental physicists (Galison 1997: 733). In conversations with me, empiricists compared the insufficient empirical basis of some modeling efforts to "trying to drive a car without the wheels" and wanting to "place the roof on a house without walls." Echoing their "extremely skeptical" fellow research meteorologists in the decades after the Second World War (Harper 2003: 675), research meteorologists I came across stressed that insufficient physical understanding structured the models. Meteorological empiricists and theoreticians enjoy insight into weather and climate dynamics that informs their critical views of the models and of how the models are both frequently used to perform uninsightful science and wrongly looked to as if they were truth machines rather than heuristic tools. They recognize that many modelers are good mathematicians but portray them, as one of them put it, as "so involved with running their models that they haven't put the time in thinking how the atmosphere works."

Some modelers acknowledge a certain factual basis in some of these criticisms, noting a common inability or reluctance among modelers to recognize their models' shortcomings (Lahsen 2005a). Aside from subcultural dynamics, the associated tensions highlight an epistemological issue characteristic of twentieth-century science: the question of whether the best understanding of the atmosphere is gathered by "those who crunch the numbers, but never look outside" or by those who do not use equations but who "read the sky" (Harper 2003).

Indicative of the entrenchment that marks scientists' positions on the climate issue, William ("Bill") Gray, a climatologist I interviewed in the 1990s, has not changed his views much in subsequent decades, at least in regard to climate change and GCM-based results. In an opinion piece he published in a newspaper in December 2010 (Gray 2010), he persists in his questioning of both. In line with his conversations with me some fifteen years earlier, Gray's opinion piece stresses the knowledge he has accumulated after a long career studying, forecasting, and teaching "meteorology-climate," and he rejects the credibility of model-based projections of temperature changes on the order of two to five degrees Celsius, describing GCMs' representation of hydraulic dynamics as being deeply flawed and producing "grossly unrealistic high warming numbers" (ibid.).

Reminiscent of Ellsaesser's isolation in his empty trailer at Lawrence Livermore, Gray's opinion piece expresses the exclusion that he and other research meteorologists feel. He writes, "Thousands of our country's older and more experienced meteorologists have similar opinions as mine," describing them all as "knowledgeable specialists" whose opinions "have yet to be included in broad, open, and honest scientific debate of the likely influence on climate by rising levels of carbon dioxide" (Gray 2010).

In another interview, a meteorologist of similar profile noted weather researchers' sense of marginalization because the IPCC, rather than the Working Group for Climate Change Detection unit (which their leaders had created in the mid- to late 1980s under the World Meteorological Organization), became the authority on climate change. Adding to the sting, members of the broader atmospheric research community discredited the unit's authority on the climate issue on the grounds that it was made up of bureaucrats and weather forecasters.[6] With an element of humor, a climatologist I interviewed caricatured the feeling he and his fellow synoptically trained colleagues have in reaction to their changed status after the rise of the numerical techniques: "Nobody loves [us] anymore and nobody knows the real atmosphere."

For their specific, experience-based reasons, but similar to the contrarian physicist trio, earlier generation climatologists' criticisms of climate models and their reservations about the current level of preoccupation with climate change are thus often infused with a sense of alienation and regret about broad-based transformations in science and society since the Cold War. They have witnessed increased internal strife and politicization of their field with the rise of environmental concern in the context of nuclear technology, global cooling, ozone depletion, and, more recently, climate change. The climatologists I interviewed were all environmentally concerned but questioned the strong focus on climate change in science and society. They often identified other issues that they thought merited more preoccupation, such as population growth and depletion of natural resources. But they were—mostly quietly so—uncomfortable about recent trends in science, including climate models and the IPCC. With characteristic ambivalence, they closely linked climate modeling with a lamented politicization of their field, while also acknowledging the power of the models as heuristics tools. What they reacted to, above all, are the uses of the models as truth machines.

As in the case of the physicist subset of contrarians, members of this previous generation of climatologists express traditional scientific values. For example, one meteorologist expressed being troubled by the IPCC because its mode of operation diverges from "the traditional role of science," according to which hypotheses are rigorously tested: "The IPCC doesn't aggressively seek to disprove its own hypothesis. The thrust of the IPCC is to look for the social and political consensus. I find that really troubling. It's really different."

Like the contrarians, skeptically inclined mainstream meteorologists I interviewed often singled out Schneider in their criticism of new trends in science. One of them described public statements he has made as "a sequence of 'ifs' and 'coulds': 'if this and if that . . . could lead to . . .' etc.," including his subsequent qualifying statement that the consequences of the hypothesized event could have either beneficial or dire consequences. This climatologist criticized this approach as lacking content and scientific rigor, commenting, "There is nothing there, right?" Reflecting a conception of scientists as being scientifically and morally superior when staying clear ("clean") of policy and politics and underscoring again the generational dimension of the GCM criticisms, he said, "[Schneider] always puts a spin on things. And who listens to him? The kids who are going to get environmental science degrees and who are going to need jobs, they listen. The ones who want to save the planet, they want to be part of this. Politicians listen . . . these scientists are their support staff. . . . But you just don't do the if-if-if, if you want to be clean. The bottom line is how well your science compares to reality."[7]

Adding insult to injury, nonmodeling meteorologists frequently found their access to research funds greatly reduced, if not entirely cut, with the emergence of GCMs and concern about climate change, especially when their research proposals were not designed to confirm or otherwise advance the theory of anthropogenic climate change. For purposes of forecasting, the synoptic methods were replaced by numerical weather forecasting, which was celebrated for its superiority. Inversely, the synoptic methods were demoted and given a new label, *subjective*, which slighted them as being less scientific. The synoptically trained climatologists had to acknowledge that the computers were better in important respects, and they even appreciate how computers have made some of their tedious former tasks unnecessary, particularly the practice of manually plotting data onto weather maps. They are also in awe of the computer technology that allows them to simulate the dynamics of the weather events they research, to examine the simulated phenomena down to the minutest details, and to slow the dynamics down in time or speed them up.

Their amazement at computers, however reticent, and the fact that some space and need for their expertise in the interpretation of model output remained soften their critiques. With a few exceptions, the climatologists I encountered during my fieldwork have largely stopped fighting the changes that have come with the new times, though they quietly continue to harbor ambivalent feelings about how environmentalism has changed their field and how sophisticated computerized methods miss some insights into complexities and dynamics. They think their scientifically trained human minds and human intuition are better able to grasp such insights by drawing not only on known laws of physics but also, centrally, on experience and a *feeling* for how the atmosphere works.

Weather Forecasters

Weather forecasters are different from the older generation of research meteorologists in that they generally do not have doctorates. Present-day weather forecasters' professional training and practice—and, thus, their perspective and subculture—overlap partly with those of the above-described older empirical research meteorologists (hereafter I refer to them as empirical meteorologists). One connecting point is their training in synoptic methods of weather analysis and forecasting. A climatologist noted that those who learn weather forecasting techniques pick up an associated subculture, even if they worked only with the techniques of doing forecasts early on in their lives and for relatively brief periods. He explained that the methods and practice of weather forecasting engender a humility and skepticism about models because of the difficulty of producing accurate forecasts. He and others noted that modelers fail to be similarly humbled when producing forecasts for climates set in the distant future; these forecasts are validated by "hindcasts," but resemblances with observations can also be an artifact of superficial adjustments ("tuning") to make models resemble past and present climates (Lahsen 2005a).

Skepticism with regard to the theory of climate change is "more the rule than the exception" among television meteorologists and weathercasters (Dawson 2008). A survey of nearly six hundred broadcast meteorologists found that only half of them believe that global warming is happening, and an even smaller subset believes that it is anthropogenic rather than due to natural variability (Wilson 2009). The survey results captured widespread attention, reinforced by recognition of broadcast meteorologists' broad influence on public opinion.

Culturally loaded tensions and resentments as well as expert knowledge appear to structure weather forecasters' positions as well. Their expertise on climate issues is often dismissed by academic scientists, who explain that weather forecasters' skepticism reflects personal bias and ignorance rather than being evidence-based.[8] Yet forecasters' formal training now includes such subjects as the IPCC and long-range climate projections (Dawson 2008), and forecasters are quite knowledgeable about data that are used in the climate models.

A senior educator of broadcast forecasters at Penn State University described the disagreements between forecasters and climate scientists as "a jurisdictional war." He mentioned the disdain in "the orthodox scientific research community" for "those who are not smart enough to get a PhD or do research, and instead go into the fluff of television and just forecast the weather." Inversely, he identified a disdain among television meteorologists "for those who pontificate about what their [climate] models show" (Dawson

2008). A broadcast meteorologist related the skepticism of "quite a few" colleagues to their experiences of having "asked questions and raised issues and been told to be quiet, [and to accept that climate change] is the truth." He said weather forecasters may resist such pressures because they "know things change that don't necessarily have to do with global warming," such as the fact that the location of certain sensors has changed, compromising the consistency of data used to detect climate trends (ibid.).

THEORETICAL METEOROLOGISTS

Theoretical meteorologists, or dynamicists, were the third group to appear in my research as mainstream scientists inclined to question GCM output. This subgroup also highlighted generational disparities and identified with those of the older school who had been trying for many years to formulate a conceptual model of how the climate system works. These efforts all took place before, in the words of one of them, "the modelers came along and said, 'It's hopeless to do it that way. We're just going to have to simulate rather than understand.' "

One mainstream dynamicist whom I interviewed in 1995 contrasted himself to modelers and to those whom he called catastrophists and the "CO_2 folks." He explicitly identified with a "group of critics," with physicists, and with "the school that believes that the climate is an extraordinarily stable system." He contested the right of anyone to portray the model output as reliable "when all it takes into account is one factor among many others" and omits important systemic feedbacks. He perceived a warming trend in global data but judged it to be roughly half a degree and thus far below other estimates.

Dynamicists have been less personally and less negatively impacted by the rise of GCMs than empirical meteorologists. They largely continue to benefit from the prestige that theoretical knowledge endows in science, and their criticisms tend to express a subtle sense of superiority in relation to the GCM enterprise, which they associate with engineering and criticize for its weak grounding in physics-based understanding of the phenomena being modeled. In a similar expression of superiority, dynamicists I interviewed sometimes expressed feeling ashamed by IPCC practices and by the climate change "hype" in their field, as the contrarian dynamicist Richard Lindzen did in *The Greenhouse Conspiracy*.

Demonstrating the moderation and complexity characteristic of mainstream skeptics, mainstream theoretical meteorologists were measured in their environmental skepticism and did not categorically dismiss a longer list of environmental issues. Indeed, some made a point of expressing disagreement with the "wide anti-science movement," which they identified with the Republican Party. While the above-mentioned mainstream dynamicist questioned contemporary environmentalist fears of nuclear technology, for example, he agreed that it makes sense to reduce CO_2 emissions for reasons unrelated to climate change.

HISTORICAL MEMORY

Historical memory is a connecting point between the contrarian and mainstream critics. Older members of both camps have witnessed model promoters' past tendencies to make unwarranted claims on the basis of uncertain model results—claims later discredited upon reevaluation and refinement of the models, techniques, and assumptions employed. A case in point was model results associated with the TTAPS study of the environmental consequences of nuclear war, a study published in 1983 and named after the first letter of each of the authors' last names (Turco, Toon, Ackerman, Pollack and Sagan 1984). The GCMs predicted a thirty-five-degree Celsius drop in temperature after the use of nuclear weapons. A strong defender of this result, the astronomer Carl Sagan, said, "I do not think that our results are dependent on some quirk internal to the computer program" (Ehrlich et al. 1984), and the biologist Paul R. Ehrlich expressed "a great deal of confidence" in them (Ehrlich et al. 1984). Using a newer, more complex dynamic model, Schneider and his colleague Starley Thompson (1988) later reduced the projected temperature change, coining the term *nuclear fall* to replace the prediction of a *nuclear winter*.

These allegations led the Atomic Energy Commission to investigate the existence of possible links between hydrogen and nuclear bombs and changes in local weather patterns. In this way, research agendas which developed along with nuclear bombs led to a focus on CO_2 through research on the trajectory of CO_2 in the climate system. Bombs and climate research are also conceptually linked in that the simulation of both involves nonlinear problems of fluid dynamics. Indeed, some of the scientists who helped conceptualize and build the simulations of hydrogen bombs later played a part in developing the numerical models by using the same methods and tools, in other words, mathematical codes and supercomputers.

The nuclear winter debate and its downgrading to nuclear fall happened between 1986 and 1988, increasing the skepticism about GCMs in some quarters immediately before a climate modeler took concerns about climate change to new heights when testifying before Congress that it already was responsible for damaging climate extremes. Personally invested in nuclear technology, the Marshall Institute physicists had particular reasons for resenting critics of nuclear technology. They were predisposed to oppose the projections of future climate change from greenhouse gas emissions because they were derived from the same models and by some of the same scientists who promoted the idea that nuclear technology potentially could cause climate change, an idea they also rejected. They saw nuclear technology as a positive contribution to humanity; indeed, Seitz was named the chair of Scientists and Engineers for Secure Energy in the 1990s, a scientific think tank funded by the nuclear industry to promote nuclear power (Deal 1993).

The nuclear winter episode reinforced skepticism among climatologists and dynamicists as well. In interviews with me, some, explicitly mentioning the TTAPS study, expressed having grown skeptical after hearing the continuous but changing "scare of the month" emanating over the years from modelers, chemists, and their cohorts.

To conclude, this analysis identifies both differences and continuities between contrarians and mainstream scientists who are skeptically inclined. Their staunch conservatism aside, contrarian scientists articulate strongly culture-laden criticisms of transformations in the sciences associated with a new age, an age characterized by rising environmentalism and a new mode of knowledge production and associated values. For reasons rooted in the particularities of their knowledge, experience, and subcultures, older generation members of three subgroups—dynamicists, climatologists, and weather forecasters—are especially inclined to question elements of climate science, namely, climate models. These three groups traditionally made up the field of meteorology and have a noncommitted—and sometimes a somewhat alienated and antagonistic—relationship to modeling, and this attitude underpins their skepticism toward climate change. Such mistrust has facilitated backlash instigators' efforts to enroll some of them in their campaigns, a few as full-fledged contrarians; the makers of *The Greenhouse Conspiracy* were able to obtain footage of nonpoliticized mainstream skeptics criticizing climate science because they tapped into tensions that exist within the scientific mainstream.

It is noteworthy that contrarian scientists tend to be empiricists and physicists, that is, theoreticians. The empiricists include Robert Balling and

Sherwood Idso, among others, and the Marshall Institute trio were physicists, as are Sallie Baliunas and Willie Soon. Lindzen is a theoretical meteorologist, while Patrick Michaels is an observational meteorologist.

Confirming the cultural theory framework, analyses of high-profile American contrarian scientists point to the crucial role of their individualistic and conservative values in their choice to join the anti-environmental movement. In their case and that of largely ignored and understudied subgroups of skeptically inclined mainstream scientists, however, other cultural dynamics are also at work. Illuminating the extent to which the particularities of scientists' training, experiences, and historical memories account for their skepticism, this chapter has illustrated the usefulness of also integrating attention to the differentiated conditions and experiences of modernity and critical analysis of climate knowledge itself. Both offer insight into the politics of global environmental change. Analysis of scientists' differentiated experiences tied to the rise of climate modeling and concern about climate change shows experience-based factors affecting perceptions of climate change, factors that cultural theory thus far has not taken sufficiently into account. This empirical analysis confirms that understanding of the cultural underpinnings of divergent perceptions of climate change requires analyses that integrate, but also move beyond, the focus on individualistic, hierarchical, and egalitarian values that characterize applications of the cultural theory framework.

The knowledge and experiences, including the scientific training, of the identified scientific subgroups structure their perceptions of climate modeling and climate change. The scientific protagonists in the climate debate have been shaped, albeit differently, by various historical moments. The nonmodeling, older generation climatologists are impacted by a sense of modernity's characteristic "maelstrom of perpetual disintegration and renewal, of struggle and contradiction, of ambiguity and anguish" that threatens to destroy everything that is known, everything that has been known, believed, and held dear (Berman 1988: 15). Their awe at what computers can do with such detail, precision, and speed is tempered not only by an equally vivid sense of what the models cannot do as well as the intuitive human mind conditioned by close empirical experience of weather dynamics; but also by the demotion they suffered now that, as one of my climatologist informants put it, "nobody loves [us] anymore."

Further analysis is needed to establish the existence of contrarian scientists outside of the United States and Europe, including the extent to which such scientists have similar or different profiles from the ones identified here. For example, as readers will appreciate after reading this chapter, it is significant that the Brazilian contrarian who wrote the *Folha de São Paulo* article supportive of the contrarian NIPCC report himself was a scientist who obtained his doctorate in physics at MIT and who, like Seitz and Singer, was of an older generation, very conservative politically, and closely aligned with the military dictatorship that ended in 1985 in Brazil.

Further analysis is also needed to establish the size of the skeptically inclined scientific subgroups and to confirm the patterns in their scientific backgrounds that I have identified here. The mainstream skeptics may not be very numerous, and they may be decreasingly active in science by reason of their age. The powerful vested financial interests and political elites behind the efforts to counter concern and policy about climate change have been clearly documented, as has their reliance on a handful or so of contrarian scientists. Such analyses have not accounted for the more moderate skepticism that persists within the mainstream scientific community, however; the latter tend to be invisible in analyses of the anti-environmental backlash, and for good reason: they are not anti-environmental. For that reason, though, it is difficult to know how many mainstream scientists would self-identify as skeptics or how many share parts of the views I encountered among a small subset of mainstream scientists. Many of the mainstream skeptics I met through my fieldwork were near or past retirement age, and nonmodeling research meteorologists with backgrounds in synoptic methods are rare among newer generations of scientists. Nowadays, the synoptic methods are taught mainly to weather forecasters and rarely at the doctoral level. Erasing the traditional division between theoreticians, empiricists, and modelers, most everyone receiving degrees in atmospheric science today ends up doing some level of modeling and is taught more theoretical knowledge than previous generations of climatologists. This could mean that the subcultural divisions between modelers, theoreticians, and empiricists may gradually disappear and possibly, with them, the associated antagonisms that lie behind the disagreements about contemporary climate science I have discussed. Since weather forecasters were not a focus of my fieldwork, whether skepticism among them implies a generational dimension is unclear.

Advocates of concern over climate change are understandably likely to welcome the disappearance of the science-focused divisions, considering how they contribute to policy gridlock. Yet mainstream skeptics' critical perspectives may carry a potential benefit. To some degree, skeptics' views reflect expertise- and experience-driven insights and highlight a "less rosy underbelly" of the climate science enterprise that IPCC leaders and advocates are unlikely to acknowledge (the emails released in association with so-called Climategate reveal such reluctance). Hence societies may be well served by some level of diversity of opinion beyond that which exists within the IPCC, in line with democratic theory (Brown 2009) and the anthropological principle of societal resilience through diversity of perspectives (Schwarz and Thompson 1990).

Backlash actors invariably highlight scientific uncertainty to intimate that climate change is less of a threat than commonly thought, but the opposite can also be the case. Thus, assumptions built into the data-collection system prevented earlier detection of stratospheric ozone depletion (Litfin 1994), and a preference for precision led the authors of a chapter in the IPCC Fourth Assessment Report to entirely omit the consequences of the rapid disintegration of the ice sheets of West Antarctic and Greenland in their estimates of sea level rise (Oppenheimer et al. 2007; O'Reilly, this volume). Similarly, at least one climatologist whom Anderegg et al. (2010) considered a skeptic contests the focus on climate change because he thinks it diverts attention from other environmental threats. Skeptical views can be socially beneficial to the extent that they sharpen our understanding of uncertain risks and avoid the dishonest manipulations and ideological inflexibility that generally mark backlash actors and their campaigns.

NOTES

I am grateful to Kristine C. Harper for helpful comments on this manuscript and sincerely thank the scientists who gave me of their time, agreeing to be interviewed and to share their perspectives and experiences with me.

1. For discussion of the complex relationship between models and observational data, see Paul Edwards (1999, 2010).
2. http://www.eecg.utoronto.ca/~prall/climate/skeptic_authors_table_by_clim.html.
3. Pierre Bourdieu's notion of habitus refers to habits and inclinations manifested in thoughts and behavior, including historical memory. Individuals' habitus are shaped by situated experiences of social processes and physical structures, which engender a sense of what is appropriate in particular circumstances and what is not.
4. The distinction between pure (or basic) science and applied science is context-dependent in the first place and often vague. In reality, much science, including climate modeling, can be considered simultaneously pure and applied (see Stokes 1997).
5. For evidence of the depth of their anticommunism and fear of military threats associated with the USSR and communism more generally, see Oreskes and Conway (2010).

6. Interview with research meteorologist Chester Ropelewski, October 29, 1999.
7. This scientist was of the same age and generation as Schneider. The generational aspect appears here in his reference to Schneider's influence on "kids" and environmentalists.
8. For examples, see Dawson (2008), in particular the comments by Bob Ryan, past president of the American Meteorological Society. Among other things, he said that some weather forecasters are "putting their own personal views—sometimes even fundamentalist religious beliefs—first, and then looking at climate change from the standpoint of preconceived things they believe in."

REFERENCES CITED

Anderegg, William R. L., James W. Prall, Jacob Harold, and Stephen H. Schneider. 2010. "Expert Credibility in Climate Change." *Proceedings of the National Academy of Science* DOI 10.1073/pnas.1003187107 (21 June).

Azevedo, J. C. de. 2008. "Qual Temperatura?" *Folha de São Paulo*, October 13.

Berman, M. 1988. *All That Is Solid Melts Into Air: The Experience of Modernity.* 2nd ed. New York: Penguin Books.

Bourdieu, P. 1991. *Language and Symbolic Power.* Edited by John B. Thompson and translated by Gino Raymond and Matthew Adamson. Cambridge: Polity Press.

Brown, M. B. 2009. *Science in Democracy: Expertise, Institutions, and Representation.* Cambridge: MIT Press.

Dawson, B. 2008. "Why Are So Many TV Meteorologists and Weathercasters Climate 'Skeptics'?" The Yale Forum on Climate Change and the Media. Last modified June 12, 2008. <http://www.yaleclimatemediaforum.org/2008/06/why-are-so-many-tv-meteorologists-and-weathercasters-climate-skeptics/>.

Deal, C. 1993. *The Greenpeace Guide to Anti-Environmental Organizations.* Berkeley: Odonian Press.

Demeritt, D. 2009. "Geography and the Promise of Integrative Environmental Research." *Geoforum* 401: 127–29.

———. 2011. "Book Review Symposium: Commentary 1." *Progress in Human Geography* 35, no. 1: 132–34.

Douglas, M., D. Gasper, S. Ney, M. Thompson, R. Hardin, S. Hargreaves-Heap, S. Islam, D. Jodelet, H. Karmasin, C. de Leonardis, J. Mehta, S. Prakash, Z. Sokelewitz, and T.-d. Truong. 1998. "Human Needs and Wants." In *Human Choice and Climate Change*, ed. S. Rayner and E. L. Malone, 1:195–263. Columbus: Battelle Press.

Douglas, M., and A. Wildavsky. 1984. *Risk and Culture.* Berkeley: University of California Press.

Duraiappah, A. 2010. "Social Sciences in Intergovernmental Processes." Presentation at Changing Nature, Changing Science? Symposium, Nagoya, Japan, December 13.

Edwards, P. N. 1999. "Data-Laden Models, Model-Filtered Data: Uncertainty and Politics in Global Climate Science." *Science as Culture* 8, no. 4: 437–72.

———. 2010. *A Vast Machine: Computer Models, Climate Data, and the Politics of Global Warming.* Cambridge: MIT Press.

Ehrlich, P. R., C. Sagan, D. Kennedy, and W. O. Roberts. 1984. *The Cold and the Dark: The World After Nuclear War.* New York: W. W. Norton.

Ellsaesser, H. W. 1989. "Response to Kellogg's Paper." In *Global Climate Change: Human and Natural Influences*, ed. S. F. Singer, 67–80. New York: Paragon House.

———, ed. 1992. *Global 2000 Revisited: Mankind's Impact on Spaceship Earth.* New York: Paragon House.

Forman, P. 1987. "Behind Quantum Electronics: National Security as a Basis for Physical Research in the United States, 1940–1960." *Historical Studies in the Physical and Biological Sciences* 18, no. 1: 149–229.

Galison, P. 1997. *Image and Logic: A Material Culture of Microphysics.* Chicago: University of Chicago Press.

Gibbons, M., C. Limoges, H. Nowotny, S. Schwartzman, P. Scott, and M. Trow. 1994. *The New Production of Knowledge: The Dynamics of Science and Research in Contemporary Societies.* Thousand Oaks, Calif.: Sage Publications.

Gray, W. 2010. "Mainstream Media Helps to Brainwash." *The Coloradoan*, December 21. Accessed October 19, 2013. http://icecap.us/index.php/go/new-and-cool/P342.

Harper, K. C. 2003. "Research from the Boundary Layer: Civilian Leadership, Military Funding and the Development of Numerical Weather Prediction (1946–55)." *Social Studies of Science* 333: 667.

Hulme, M. 2008. "Geographical Work at the Boundaries of Climate Change." *Transactions of the Institute of British Geographers* 33, no. 1: 5–11.

———. 2009. *Why We Disagree About Climate Change: Understanding Controversy, Inaction and Opportunity.* Cambridge: Cambridge University Press.

Jasanoff, S. 2010. "A New Climate for Society." *Theory, Culture and Society* 27 (March): 233–53.

Kahan, D. 2010. "Fixing the Communications Failure." *Nature* 463 (January): 296.

Kevles, D. 1990. "Cold War and Hot Physics: Science, Security, and the American State, 1945–56." *Historical Studies in the Physical and Biological Sciences* 20, part 2: 239–64.

Kraft, M. E. 2000. "Environmental Policy in Congress: From Consensus to Gridlock." In *Environmental Policy*, 4th ed., ed. N. J. Vig and M. E. Kraft, 121–44. Washington: Sage Publications.

Lahsen, M. 1998. "Climate Rhetoric: Constructions of Climate Science in the Age of Environmentalism." PhD diss., Rice University. UMI Dissertation Services.

———. 1999. "The Detection and Attribution of Conspiracies: The Controversy Over Chapter 8." In *Paranoia Within Reason: A Casebook on Conspiracy as Explanation. Late Editions 6, Cultural Studies for the End of the Century*, ed. G. E. Marcus, 111–36. Chicago: University of Chicago Press.

———. 2005a. "Seductive Simulations: Uncertainty Distribution Around Climate Models." *Social Studies of Science* 35: 895–922.

———. 2005b. "Technocracy, Democracy and U.S. Climate Science Politics: The Need for Demarcations." *Science, Technology and Human Values* 30, no. 1: 137–69.

———. 2008. "Experiences of Modernity in the Greenhouse: A Cultural Analysis of a Physicist 'Trio' Supporting the Conservative Backlash Against Global Warming." *Global Environmental Change* 18, no. 1: 204–19.

———. 2010. "The Social Status of Climate Change Knowledge." *WIREs Climate Change* 1, no. 2: 162–71.

———. 2013. "Anatomy of Dissent: A Cultural Analysis of Climate Skepticism." *American Behavioral Scientist* 57, no. 6: 732–53.

Lawler, A. 1995. "NASA Mission Gets Down to Earth." *Science* 269 (September 1): 1208–10.

Leiserowitz, A., E. Maibach, and C Roser-Renouf. 2010. *Climate Change in the American Mind: Americans' Global Warming Beliefs and Attitudes in January 2010.* Yale Project on Climate Change. Yale University and George Mason University. New Haven: http://environment.yale.edu/uploads/AmericansGlobalWarmingBeliefs2010.pdf.

———. 2013. Yale Project on Climate Change. Yale University and George Mason University. Accessed October 19, 2013. http://environment.yale.edu/climate-communication.

Leiserowitz, A., E. Maibach, C. Roser-Renouf, N. Smith, and E. Dawson. 2013. "Climategate, Public Opinion, and the Loss of Trust." *American Behavioral Scientist* 57 (June): 818–37.

Litfin, K. T. 1994. *Ozone Discourses: Science and Politics in Global Environmental Cooperation.* New York: Columbia University Press.

Lowry, W. R. 2000. "Natural Resource Policies in the Twenty-First Century." In *Environmental Policy,* 4th ed., ed. N. J. Vig and M. E. Kraft, 303–25. Washington: Sage Publications.

Martin, E. 1998. "Anthropology and the Cultural Study of Science." *Science, Technology, and Human Values* 23, no. 1 (Winter): 24–44.

McCright, A. M., and R. E. Dunlap. 2000. "Challenging Global Warming as a Social Problem: An Analysis of the Conservative Movement's Counter-Claims." *Social Problems* 47, no. 4: 499–522.

———. 2003. "Defeating Kyoto: The Conservative Movement's Impact on U.S. Climate Change Policy." *Social Problems* 50, no. 3: 348–73.

Mouffe, C. 1988. "Hegemony and New Political Subjects." In *Marxism and the Interpretation of Culture,* ed. C. Nelson and L. Grossberg, 89–101. Urbana: Board of Trustees of the University of Illinois.

National Research Council. 1998. *Capacity of U.S. Climate Modeling to Support Climate Change Assessment Activities.* Washington: National Academy Press.

Oppenheimer, M., C. O'Neill, and S. Agrawala. 2007. "The Limits of Consensus." *Science:* 1505–6.

Oreskes, N., and E. M. Conway. 2010. *Merchants of Doubt: How a Handful of Scientists Obscured the Truth on Issues from Tobacco Smoke to Global Warming.* New York: Bloomsbury Press.

Pendergraft, C. A. 1998. "Human Dimensions of Climate Change: Cultural Theory and Collective Action." *Climatic Change* 39: 643–63.

Proctor, J. D. 1998. "The Meaning of Global Environmental Change: Retheorizing Culture in Human Dimensions Research." *Global Environmental Change* 8, no. 3: 227–48.

Schneider, S. H., and S. L. Thompson. 1988. "Simulating the Climatic Effects of Nuclear War." *Nature* 333 (May 19): 221–27.

Schwarz, M., and M. Thompson. 1990. *Divided We Stand: Redefining Politics, Technology and Social Choice.* Hemel Hempstead: University of Pennsylvania Press.

Seitz, F. 1994. *On the Frontier: My Life in Science.* New York: American Institute of Physics.

Silverman, H. 2010. "Climate, Worldviews and Cultural Theory." *People and Place* 1, no. 2 (March 9): http://www.peopleandplace.net/media_library/image/2010/3/9/climate_worldviews_and_cultural_theory.

Singer, S. F., ed. 2008. "Nature, Not Human Activity, Rules the Climate: Summary for Policymakers of the Report of the Nongovernmental International Panel on Climate Change." Chicago: Heartland Institute.

Stokes, D. E. 1997. *Pasteur's Quadrant: Basic Science and Technological Innovation.* Washington: Brookings Institution Press.

Thompson, M., and S. Rayner. 1998. "Cultural Discourses." In *Human Choice and Climate Change*, ed. S. Rayner and E. L. Malone, 1:265–343. Columbus: Battelle Press.

Turco, R. P., O. Toon, T. Ackerman, J. Pollack, and C. Sagan. 1984. "Nuclear Winter." *Science* (December 23): 1284.

Wilson, K. 2009. "Opportunities and Obstacles for Television Weathercasters to Report on Climate Change." *Bulletin of the American Meteorological Society* 90 (October): 1457–65.

Chapter 10 The Uniqueness of the Everyday: Herders and Invasive Species in India

Rajindra K. Puri

> Every day, a man in the heart of Borneo leaves his home early in the morning and walks into the forest, dogs following, spear in hand.
>
> Every day, an elderly woman in the Pantanal of Brazil rises early and heads to the river's edge with her lines and hooks in hand.
>
> Every day, a young boy in India opens his corral early in the morning and follows his cattle into the forest, bill knife in hand.

All of these scenarios describe the opening acts of hunting, fishing, and herding activities that take place every day, in similar forms, in myriad communities around the world. But while the everyday conjures a sense of the mundane and routine, in these apparently widely differing scenarios every day is in fact unpredictable: what happens next, nobody can know. Filled with risks, uncertainty, mystery, and probable contingency, their unfolding and outcomes are unpredictable rather than routine. This potential for drama, eventual success or failure, makes these everyday activities exhilarating and compelling to follow, always memorable and, most important in the context of this chapter, revealing.

Yet in spite of the unpredictability, people everywhere do succeed, often with surprising regularity, in catching wild pigs and fish and in bringing their cows home safely. This success is usually attributed, by outsiders, to the development over time of substantial bodies of knowledge about all aspects of the environment, such as weather, prey, tools, skills, and places (Puri 2005). The participants themselves may point to such factors as their knowledge, the right physical, social, and even metaphysical environmental conditions, fate or luck, proper moral comportment, and attentiveness to ritual responsibilities (Garay-Barayazarra and Puri 2011; Salick and Ross 2009).

While there are real similarities between these scenarios in terms of their unpredictability and how actors adapt to the uncertainties presented to them, the commonality I want to discuss here is one that has only recently come to connect these distant locations, namely, climate change and its impacts. Changes in temperature, rainfall, sunshine, and winds threaten to increase or decrease known risks associated with these daily life-giving activities, possibly presenting surprises that are beyond the scope of local knowledge systems. As an ethnobiologist, I am curious about how environmental knowledge may be acquired, discarded, borrowed, innovated, and transformed in order to maintain success in provisioning families. As an anthropologist, thinking holistically about such circumstances, I add that the direct and indirect impacts associated with climate change are bound to change the total context in which these events take place. Such a context might include changes in biodiversity, water and soils, other subsistence activities such as agriculture, household livelihoods, kin and social structures, cosmologies, values, belief and ritual systems, local economies and markets, regional demographics, institutions and governance at all scales, development and conservation policies, and the discourse of climate change and its impacts. All of these and probably many more are necessary elements in a conceptual framework that will help us to understand how and why these everyday activities are proceeding, whether they are changing, and what kinds of outcomes they will lead to. The value of these outcomes is dependent on the particular local actors, families, communities, and nation-states in question.

Because of their vulnerability to climate change as biodiversity-dependent people there is merit in considering the ways in which hunters, fishers, and herders deal with uncertainty. Furthermore, as experts in contingency they can serve as models not only for studying and thinking about the ways in which all people will respond to increased uncertainty associated with global environmental change but also for understanding how those responses will be not so much the direct result of climate change but the result of changes in the wider social and cultural context. I propose an anthropological approach to understanding the impacts of and responses to climate change better: to follow, to watch the unfolding of what are familiar activities, now in new contexts, into the unknowable future; to see what happens, what changes, and how people respond, and then, using whatever tools and methods are relevant, to try and understand why these events have occurred as they have and led to the outcomes they have.

This approach to understanding climate change and its impacts through a focus on the quotidian activities that constitute the basis of people's lives draws inspiration from two main sources: Tim Ingold's writings on dwelling and wayfaring (2000, 2007, 2011), and A. Peter Vayda's and Bradley Walters's methodology for explaining changes in human–environment relationships by means of causal analyses of events, which they call abductive causal eventism (Vayda and Walters 2011; Walters 2012). These seemingly different approaches in fact share the underlying theoretical assumption that context, both temporal and spatial, is critically important to understanding human decision making in the face of change.

Ingold's work implies a need to focus closely on fine-grained aspects of daily life to understand more fully how people actually live with the world, which seems precisely attuned to finding out how people are changing into the future. Ingold writes, "Here lies the essence of what it means to dwell. It is, literally to be embarked upon a movement along a way of life. The perceiver-producer is thus a wayfarer, and the mode of production is itself a trail blazed or a path followed. Along such paths lives are lived, skills developed, observations made and understandings grown. . . . The path, and not the place, is the primary condition of being, or rather of becoming" (2011: 12). He continues, "Whether our concern is to inhabit the world or to study it—and at root these are the same, since all inhabitants are students and all students inhabitants—our task is not to take stock of its contents but to *follow what is going on*, tracing the multiple trails of becoming, wherever they lead" (ibid.: 14). Ingold's phenomenological approach can describe the dynamics of life lived in changing contexts, and specifically the instances, activities, and events that might be identified as potential responses to such causal factors as climate change. Analyzing these responses involves using Vayda's and Walters's methodology to determine their causes, honing in on an explanation for those responses by reconstructing a plausible history of the events that led to them and along the way disproving hypothesized causal factors, including, sometimes, those associated with climate change.

The first challenge is to identify what constitutes a response, which is more problematic than one might think (Puri 2012). The *Oxford English Dictionary* definition of *response* as "a reaction to something" (from the Latin *responsum*, "something offered in return") implies an effect-to-cause relationship, yet the tendency methodologically is to start with causes and look for effects. A melting glacier or an invasion of pests could be a readily observable event that could be considered a cause that provokes responses by people, that is, effects, such as moving one's house or buying pesticides, which might also be readily observable if one is in fact "following" life as I have suggested. But many have warned that such a strategy risks "confirmation bias," through the ignorance of alternative and multiple causes for the effects observed (Vayda and Walters 2011). The proper approach must be to identify effects, that is, responses, and then seek causes. That said, there are many challenges in identifying responses in the field, given that they may not be discrete events or readily observable, and they may be direct or indirect, individual or coordinated. In addition, responders may not be fully cognizant of all or even some of their responses (Puri 2012).

If we can identify a response, the next challenge is to analyze the myriad factors, like the ones mentioned above, that affect daily activities and will be of relevance in explaining the eventful pathways people take in living their daily lives. Maybe changes in weather or the consequences of those changes will be determinant; maybe it will be the efforts of the state itself to mitigate emissions or other hazards that drive local responses (Dove 2007), as we see in REDD+ projects (Dooley et al. 2011; Morgan 2011; cf. Danielsen et al. 2011; Mathews, this volume; McElwee, this volume); maybe it will be those forces of globalization, the wider political economy, that drive social or ecological change, and the biodiversity shifts caused by reduced rainfall and higher temperatures will just push those changes along. As Crate and Nuttall (2009: 11) note, "Climate change is a threat multiplier. It magnifies and exacerbates existing social, economic, political and environmental trends, problems, issues, tensions, and challenges."

Many anthropologists, including Roy Rappaport in *The Anthropology of Trouble* (1993), have pointed to the need to examine cosmologies, worldviews, meanings, and values—the stuff of culture we all know and love to champion—in trying to understand the actions and reactions to the ideas and physical changes brought on by climate change (Hulme 2012; O'Brien

2009; Puri 2007; Rayner 2003; Roncoli et al. 2009; Strauss and Orlove 2003). But while this is essential methodologically, maybe it will not matter too much or will only matter for some or for only some of the time (e.g., Lipset 2011). Maybe the material reality of life in changing places or new places, like refugee camps, will offer new contexts for cultural transformation. We will need to be much more open to exploring the subjects of other disciplines and to collaborating with experts in other disciplines in order to discover what linkages or combinations of linkages are actually being activated and are meaningful in explaining the events and processes we are witnessing and going through ourselves.

We should not be narrowly focused on searching for climate change impacts and responses, for that may blind us to both other events impinging on people's lives and the social and cultural contexts that are also dynamic and affect what and how people do what they do. Because of the unpredictable nature of activities in a changing context, my descriptions of the events as they occur are necessarily forward looking and move between local conceptions and practices and an outsider's observations and interpretations. Explanations for these events can range over the domains of knowledge, values, livelihoods, institutions, local and national politics, and even global discourses about progress and climate change.

I have been exploring these challenges in the course of conducting a pilot study of indigenous Soliga people's responses to the invasion of the tropical American shrub *Lantana camara L.* in Karnataka, India. The spread of this invasive species is said by the Soliga to be exacerbated by changing climatic conditions. This research was initiated as part of a multidisciplinary project called Human Adaptation to Biodiversity Change (HABC) (Howard 2009, 2012). In describing the project and discussing some of its initial findings, I hope to demonstrate both my approach and some of the above-mentioned difficulties in identifying responses, which I see as the starting point for analyses of ongoing and everyday climate change adaptation.

HUMAN ADAPTATION TO BIODIVERSITY CHANGE

Biodiversity change is expected to be one of climate change's more uncertain consequences, although the movement of plant and animal species ranges across altitudes and latitudes, and the shifting of seasonal transitions has already been observed (Parmesan and Yohe 2003). Changes in the local climate, habitat, food resources, predators, dispersers, and pollinators can be expected to combine in uncertain ways to affect species and ecosystem ranges, daily behavior, and reproductive biology, all of which may threaten or enhance the survival of species and ecosystems. Examples are beginning to mount and are making the popular news (Huffington Post 2012): early-blooming cherry trees in Japan are frustrating the annual festival planners and causing economic hardship; pine beetles are ravaging the forests of British Columbia instead of dying off with the first freeze of winter (Maness et al. 2012); in China as the frost line moves north, snails carrying the disease schistosomiasis are expected to increase their range, putting them into contact with twenty million more people (Zhou et al. 2008); in Alaska a parasite in the Yukon River is already having dramatic effects on the reproduction of salmon (Zuray et al. 2012). Such changes are bound to affect humans, compromising ecosystem services and well-being, both directly and indirectly, and leading to short-term responses and long-term strategies at all scales of human–environment interaction.

However, while we know about some of the changes to biodiversity already occurring, and many are looking at how changes in temperatures and rainfall, sea level rise, and melting glaciers and ice caps might affect settlements and agriculture (Crate and Nuttall 2009; Morton 2007; Orlove et al. 2008), a gap remains in research on how people are adapting or will adapt to biodiversity changes. The aim of the HABC project is to promote understanding of how people are responding to changes in biodiversity and related ecosystem services within societies that are highly dependent on biodiversity and have little help from state bureaucracies.[1] This project should begin to provide researchers and policy makers with tools for comprehending and supporting more thoroughly response and change from below.[2] We use invasive species as an example of biodiversity change that can alter ecosystems in unexpected ways (Cronk and Fuller 1995) and thereby mimic the expected effects of climate change on biodiversity in the future.[3] We are well aware, though, that the effects of climate change may already be impacting our study area.

Fieldwork was conducted in the Malai Mahadeshwara (MM) Hills Wildlife Sanctuary of southern Karnataka (fig. 10.1) among Soliga tribals (former forest hunters and gatherers) and Lingayat farmers, who are attempting to adapt to the invasion of the tropical American shrub *Lantana camara L.* (Verbenaceae) (fig. 10.2). *Lantana* was brought to India by British colonials in 1807 as a garden ornamental. It subsequently escaped the garden fences and has covered much of southern India (Kannan et al. 2013). One of

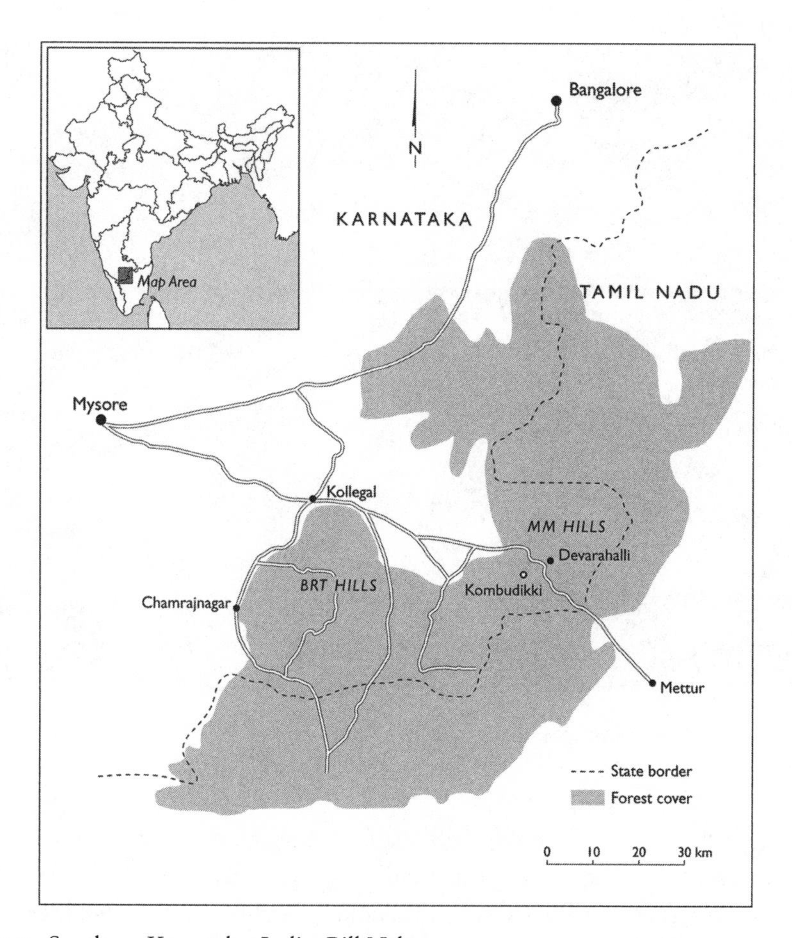

10.1: Southern Karnataka, India. Bill Nelson.

the world's top ten superinvasive plants, it has devastated as many as seventy countries, including Australia and South Africa (Bhagwat et al. 2012). *Lantana* has invaded our study area fairly recently, but by 2001 as much as 80 percent of the understory of forests in MM Hills was reportedly impacted by it (Aravind et al. 2006), with densities in dry deciduous forest, the principal vegetation in the reserve, reaching up to two thousand stems per hectare (Aravind et al. 2010)

We worked in the small village of Kombudikki, located 5 km from the main town and temple of Devarahalli in MM Hills Wildlife Sanctuary, and already the site of research being conducted by our partners at the Ashoka Trust for Research on Ecology and Environment (ATREE)[4] (Aravind et al.

10.2: *Lantana camara.* Rajindra Puri.

2006; Desai and Bhaya 2009; Uma Shankaar et al. 2004, 2010; Sharma et al. 2005). Both ethnic groups in Kombudikki—approximately thirty Soliga and eighty Lingayat households—are smallholder farmers living on holdings of between 0.5 and 2 hectares, where they grow millets and beans as food staples and some maize, cassava, castor, and sunflower as cash crops. They till the land by hand, with cattle, and, more recently, with rented tractors and may rely on family, hired, or exchange labor to do so. Households supplement their diet with leaves, vegetables, fruits, honey, and tubers from home gardens, fields, and forest, especially during the rainy season. The Soliga engage in some hunting of small game such as jungle fowl and lizards for consumption. Cash is also obtained through labor migration to quarries, sale of nontimber forest products, fuel wood, bamboo baskets, and livestock (cattle and goats). The government of India's Public Distribution System provides essential supplemental food to all households classified as being below the poverty line (Kent and Dorward 2012).[5]

The village is surrounded by rolling hills of open, dry deciduous forest and bare rocky peaks and outcrops, with species of *Acacia, Memecylon,*

Canthium, Phyllanthus, Ficus, bamboo, and some sandalwood (for the most part cut out and only beginning to regenerate). These forests are home to wild elephants, sloth bear, tiger, leopard, boar, deer, langurs, and wild jungle fowl, to mention just those fauna of import to humans. The forests have been legally owned and managed by the Indian Forest Department since the late nineteenth century, and they are patrolled periodically. The Soliga were supposedly forced to give up forest dwelling and swidden cultivation at that time and settled in hamlets in the valley.

Lantana, known locally as *jeddi gida,* "useless plant," is said to have invaded the forests of Kombudikki following the mast flowering and dieback of bamboo (*Bambusa bambos* (L.) Voss) in the early 1970s.[6] Locals helped to clear out large areas of the dead bamboo, and it was taken to the Mysore Paper Mill. The opening of the canopy and the general disturbance caused by people and trucks working in the forest is believed to have created the conditions that promoted *Lantana*'s spread (Bhagwat et al. 2012; Desai and Bhaya 2009). Residents of Kombudikki claim that bamboo harvesting was the initial catalyst but that increasing dry spells and erratic, reduced rainfall are also creating the conditions for the spread of *Lantana.* They appear to have no knowledge of global warming or of the international discourse of climate change and have no ready explanation other than natural variation.

Today, both Soliga and Lingayat communities make regular use of the forest for grazing livestock and for gathering bamboo for basket weaving (*Dendrocalamus strictus,* "kibideru"), wild foods (fruits, leaves, honey, and tubers), wild palm leaves for brooms (*Phoenix sylvestris,* "broomstick"), commercialized forest fruits (primarily *Phyllanthus emblica,* "gooseberry," and *Sapindus trifoliatus* "soapnut"), and bamboo and timber for construction. Some Lingayat farmers collect and sell firewood to middlemen in Devarahalli (Kent and Dorward 2012).

Our preliminary results indicate that the presence of *Lantana* in the forest undergrowth leads to a number of changes in biodiversity that people may be forced to respond to. The first and most obvious change arises from the presence of a new species in large numbers. Second, *Lantana* outcompetes individuals of other species, principally herbs and grasses, thus altering vegetation composition. Third, at high density *Lantana* may impact faunal biodiversity through its direct impact on food availability—its nectar and berries but not its leaves are palatable to a range of fauna—and mobility: the dense, thorny understory vegetation is impassable to larger animals. Fourth, the displacement of native flora has an effect on the fauna that use it, such as driving wild herbivores as well as their predators from the forests, thereby increasing human–wildlife conflicts. Finally, *Lantana* may have an indirect effect on species composition though its impact on soils, water, and other ecological processes (Aravind et al. 2006, 2010; Bhagwat et al. 2012; Uma Shaanker et al. 2010; Sundaram and Hiremath 2012).

In Kombudikki I focused my research on understanding the everyday forest-based activities that might be changing as a result of the invasion of *Lantana*, such as the collection of nontimber forest products like edible leafy greens, medicinal plants, fruit, honey, bamboo, and palm leaves for subsistence and sale, gathering of fuel wood, cattle grazing, and hunting. My approach was to simply follow people into the forest and observe how *Lantana* was directly and indirectly affecting what they were doing. To what extent did they interact with the plant? When a decision was made to stop, move, take a particular route, collect something, or return home, how did *Lantana* impact that decision, if at all? How did respondents explain how their practices were changing, if indeed they were, and why?

Other HABC colleagues started with households and their assets and first examined the kinds of changes that had occurred recently (Kent and Dorward 2012). If effects of *Lantana* were mentioned, then these were explored. In this way these methods were also designed to contextualize any responses to *Lantana* in terms of changes and responses occurring at the level of livelihood strategies. This context proves crucial in understanding whether someone decides to go into the forest in the first place, but the responses mentioned still need to be demonstrated and understood in the actual living of life, day to day, the variability and messiness of which tend to be smoothed out in respondents' narratives of what they do.

GRAZING CATTLE IN THE FORESTS

To Lingayat and Soliga, cattle are cultural keystone species: sacred and venerated, providing milk for consumption, essential to agriculture for plowing and their manure, bringing in some annual income through the sale of male calves, serving as collateral for loans, and representing a long-term asset to be sold for marriages, land purchases, and paying off debts. Accordingly, to many people, life would be neither desirable nor possible in Kombudikki without cattle. Yet almost everyone I spoke to agreed that keeping cows was increasingly difficult or undesirable and reported that the number of households with cattle and the total numbers of cattle had decreased in the past ten years. The causes of the decline were often attributed to shrinking household sizes or to a preference for quarry labor or both, meaning there are fewer hands around to take care of the cattle every day. Some could get around this constraint by simply paying others to do the job. Others pointed to the difficulties of feeding cattle because the *Lantana* in the forests was covering up and suppressing fodder grasses. They say this has led to underfed, malnourished animals, a circumstance which has led to their increased vulnerability to disease, injury owing to accidents, and attack by wild animals, such as leopard. In addition, people feared for their own safety in these circumstances: having to take cattle further into the forest, on to steeper and more marginal terrain, and having to stay longer every day. During our fieldwork, one resident tragically died from injuries suffered during an attack by sloth bears while grazing his cattle less than a kilometer from the village.

Typically, for most of the year cattle are taken to the forest to graze on forest grasses and herbs, their owners leaving every day after breakfast around nine o'clock and returning around five. However, it appears that if the herd numbers more than twenty head, it is more efficient just to leave the cattle in the forest at a cowshed guarded over by a family member or a paid herder, who then grazes the cattle in the surrounding forest every day. Some cowsheds are nearby, within four hours (10 km), but some can be as far as 25 km away, close to the border with Tamil Nadu. During the very hot, dry months of February to May, when forest grasses dry up and die back, most farmers bring their cattle home to stay in their harvested fields and feed on agricultural waste and stored fodder grasses harvested by hand from the fields and nearby scrub during the growing season. The few farmers with large herds, of fifty head and above, will keep them in forest cowsheds, though they may have to move them to other parts of the forest.

In questioning herders about their choices to either stay in the forest or return to the village and where to go when grazing, we were told by some that the cows themselves know where to go to find the best grazing areas: one just needs to open the gate and they will lead the way! Others said they direct the cattle to locations they know to have good grasses and let the cattle graze freely in that general area. There was no mention of formal discussions about who would go where each day or of territories, for example, for the Soliga versus Lingayat, or other tenure institutions that might affect their choices. It seemed, at first, that the forests were an open-access resource for all residents, until I remembered that the forests are owned by the state, and technically most of these activities are illegal. I was told these forest-based activities continue to happen, as they always have, because enforcement by the forest service is sporadic and bribes are paid.

To gain more insight into how grazing was affected by *Lantana* and what actually happened in the forests, I joined several herders on their day trips to the forest with cattle. Accompanying me was my research assistant and translator, Kumar Chinmayee, from ATREE. Based on field notes and transcripts, the following narrative of how one of these trips unfolded, including seemingly mundane observations and tangential events, represents both the methodology I am proposing for the investigating of adaptation to climate change impacts, examining closely the everyday activities of people, and the eventfulness and unpredictability of a day's progression. In the following analysis I highlight the difficulties of determining what aspects of these activities are responses and whether their causes are to be found in events influenced by climate change or other social or economic processes:

> 15 November 2011, 9:40 a.m.: We leave Puttamada's house and corral with his son Giri, research assistant Kumar, and 9 cattle (incl. 3 calves). Puttamada took them yesterday to Baren Bare, so Giri takes them today. The cattle lead the way, Giri urging them on, us at the back. Kumar asks him where we are going and who decides where to go. Giri claims he decides where to take the cows: "Some place three miles distant; some place where it is clean . . . a place where there is not much stones, dust, or thorns; where grass is amply available."
>
> We take the main trail into the forest, southeast toward Koti Kere junction. The trail is lined with *Lantana;* Giri tells us this is the abandoned terrace field of a Lingayat family that doesn't have the labor to clear it. We pass out of this area into the forest proper, the top of a hill, named Hangasara gudde ane, or "women's dancing place." "That is where they danced; in those days, it used to rain when they danced," Giri tells us. We've heard this before. Only women were allowed, and they'd dance when the rains were late or there was a prolonged drought.
>
> We are joined by a herd of goats, a cow, and two Lingayat people . . . they turn right, off the main trail and, we are told, head downhill toward Emmunde bare, "buffalo's rock," where a buffalo once fell and died.
>
> About 10 a.m.: At the sharpening log (*gude hane*) in the dried stream ravine (*kampayyen halla*), we spot a giant squirrel (blackish with light reddish tail). Giri says they don't hunt or eat these squirrels. This leads to a

conversation about food. Giri doesn't bring lunch and will drink water from pools in the forest if need be, "We are not used to eat lunch in the afternoon."

About 10:30: Puttamadappa (Lingayat) and his 17 cows come down the trail. Manikenta, his nephew, is not there, but will follow later. We ask him where he is going. "Koti kere," he answers. "Won't you go to Soligatti?" Kumar asks (where he went two days ago with us). "I might go, if the cattle lead me," he says. "If they don't, you go wherever they lead you?" asks Kumar. "Yes," he answers. Kumar turns to Giri and asks him why Puttamadappa follows his cows, while Giri leads his. Giri answers, "I go here; we know what all there is in this forest; we know where they [his cows] go; we also know what they do; they move like that; that is why we do what they want!"

We rest at the edge of the trail while cows graze on grasses nearby (mostly on *baale hulu* and *ulugadulu*). Giri doesn't seem to pay them much attention, just sits in the shade, while they wander through the bush eating. There are clumps of bamboo (*D. strictus*) at the edge of this part of trail. Many are broken off at about two meters. There is old dried elephant dung on the narrow trails that weave between the clumps, and parts of the more exposed areas are covered in lantana shrubs over 2 m in height. I think elephants eating here have opened up the area to *Lantana* invasion, but Giri (and others) think the *Lantana* comes anyway and inhibits bamboo shoot growth, so elephants also suffer.

Less than 10 m from the trail, the hillside is exposed, very steep, and interspersed with bare granite rock slabs, with grasses in the various crevices and cracks that transect the slabs. The cows move in a casual and somewhat broken line up the narrow trails that switch back up the slope, some veering off to the left or right to graze on a patch of grass. Giri leads the way up, and he calls out to them loudly as we climb, and they follow slowly, eating as they go. Giri seems in no rush. Both he and Kumar check their phones for a signal.

Kumar asks about loaning the bullock out for ploughing. Giri gets 300 rupees (Rs) plus lunch and *ragi* grass for the cow for a day's work ploughing in the fields. With regard to sharing cattle or labor to manage cattle, Giri claims they only take care of their own cows and their own family, not other people, even if they are Soliga, because of political affiliations: "Not even if they pay us; they are from Congress; ours is Janata Party. That is where the differences arise."

There is less *Lantana* up here, very small bushes where there are deeper pockets of soil or a lone stem in one of the narrow crevices. It is noticeably clearer and less oppressive, the large sheets of bare rock creating more space. It is precarious, though, and one can see that it would be very easy for a slip to send a person or animal off the mountainside to their death on the rocks below.

We rest under the shade of a small sapling on a rather narrow ledge about 30 m up. This area is known as Kampayn Halla ("Kampayn's ravine or dried up stream bed," actually lined with the larger *Bambusa* bamboo, *bideru*, to the west), which runs north–south on the slope below the large rocky outcrop known as Itchi bare kate ("fig rock pool"). We watch the cows grazing, some venturing out on the small grassy ledges, most sticking to the bushes and flatter bits. Giri calls out to his cattle periodically, "That sound makes them feel at ease; as if someone they know is nearby." We also speak with Giri till about noon, about animals, *Lantana*, and grazing. He says there are fewer cattle being kept by people these days, and less being grazed. "It is less nowadays; . . . previously if there were 200 houses in a village, so there would be 2,000 cows; just imagine their number if we had cows like that! It was 2,000 cows then; now there are maybe 200 cows from 10 houses." He continues, "There are not many experienced grazers, some of them are not used to grazing them; because they don't want to go any more. They don't care for the cows. Sometimes they go off suddenly, then it is a waste. Sometimes a tiger or bear catch hold of them. If they have some 10–15 cattle, they can live comfortably by selling them off. . . . If they look after them every day, instead of selling them, naturally their number would increase."

He claims the grazing is good here. In 11 years they have built up their herd from zero to 9. Before they had cows, they worked the land with hoes and moved rocks by hand. We discuss fodder grasses in the forest, and he names 9 local taxa. *Naale Hullu* [*Eragrostis burmanica* Bos.] is considered the best for cattle, it makes them strong and gives lots of milk for the calves. He claims that there has been no change in the number of grass varieties available since he was a child, they are all still here. During summer months they have to cut leaves from various trees to feed their cattle, including bamboo leaves and grasses and agricultural waste found in the village. They can feed on bamboo shoots too, but they must do so on their own. If people feed them it will be poisonous. Kumar asks him again, why come here for grazing: "There are good trees around here," he says. "We don't find them on the other side. . . . The cows graze by themselves even if we bring them here for 15 days; we can simply sleep. . . . Otherwise you have to run behind the cattle; they move helter skelter." Then he rolls over for a half-hour nap!

At 12:30 Giri wakes up and looks around, calls the cows and says we will go up further. The cattle are already up above us and, according to Giri, making their way toward the ridge. This climb is even steeper, and we have to leap between boulders and scramble up to another rocky ledge, wider, under more shade and larger trees, and on a trail that runs east to west, just underneath the rocky ridgeline and Itchi bare kate. There is more elephant dung along this trail and a few clumps of bamboo, including one that does have longer culms. There are larger shrubs of *Lantana* established in the open spaces near these clumps.

Giri sits us down and goes off to collect bamboo, just one long culm is all he needs to prepare enough bamboo to make 2 small baskets (about 20 cm in d.). He uses his bill knife for all aspects of the cutting, splitting, and shaving of both the ribs and the weaving strips. This takes about 20–30 minutes. Once he has all the pieces, it takes him another 20 minutes to make the basket. We stay in this area for about two hours, chatting with him as he makes two baskets. There are still good areas to collect bamboo, such as Andebin Halla, Emmunde Bore, Itchi bare kate. We ask why it seems only men make baskets. He responds that there are some women who do so as well. He says it's typical that they make baskets while watching the cows, "so as not to waste time" and when they go to the cowshed they can make many more. Their old cowshed at Badoddi is also a good place to collect bamboo, so when they go there later this week for ten days they will make many baskets. When grazing he may collect 5–10 or so culms to bring home, any more and that distracts them from grazing. In the past they used to collect more. They have less now because the cattle eat them.

He claims to make about 600 Rs. a week from baskets. He can get 10 Rs. for a small basket (30 a week), 25–30 Rs. for a medium basket (25 a week), and 50 Rs. for a large basket (20 per week). Baskets sell for more in MM hills (20 Rs. rather than 10Rs. for a small one). There are fewer basket makers now, only 12 Soliga men make them.

Around 3 p.m. we move westward and up toward the ridge, meeting the main trail coming from Baren Bare ("hunter's rock"). This is clearly a well-worn trail, not more than a meter wide in most places, lined with bamboo clumps, old elephant dung, larger and shadier trees. The cows are far ahead of us, having already wandered this way while we were sitting. We pause here and

> there to chat as Giri calls the cattle, grazing in the surrounding shrubs, onto the main trail. We pass through a clearing well worn by the feet of elephants, with lots of dung in and around it. According to Giri, elephants like to rest here.
>
> We come out where we started, at the Hangsara Gude Hane, the junction on the top of the ridge overlooking the village. From there we head home and are back by 4 p.m.

This proved to be a very productive day, both in terms of the observation of grazing and the information gleaned while informally chatting with Giri. What then, in all of this, constitutes a response event to biodiversity change? Is it in the route chosen and the length of time spent? or perhaps in the various stops and starts? or the accompanying activities? How do we know these events were not owing to other factors? And to what extent was the day's unfolding a response to our presence as observers?

RESPONSE EVENTS AND THEIR CAUSES

Identifying the response events that can then be analyzed using Vayda's and Walter's abductive causal eventism method is the critical first step in determining if and how *Lantana*'s presence may be affecting this livelihood activity. Close observation of a day's herding, following Ingold's approach, is absolutely necessary to begin this process. The analytical difficulty in studying adaptation in this case is in distinguishing response events that might be caused by biodiversity change from those that might be due to other events associated with other processes and also from those that might have a complex causal history driven by multiple, interacting processes, including biodiversity change.

The lack of direct interactions with *Lantana* was surprising. There was no trail clearing or cutting back of the shrub as we traveled. This has proven to be the norm in Kombudikki. Very rarely does anyone bother to cut back the *Lantana* on the trail, despite the fact that many of the trails have been narrowed considerably. As most people reported, it is pointless work, so they just "push on through," despite the irritation caused by the thorns. There was little to no attention paid to the cattle and their interactions with *Lantana* either. Again, this is surprising considering the leaves are allelopathic, but then the cows must "know" this.

In fact, I was amazed by the general lack of attention paid to the cattle at all and by the fact that they did indeed seem to be on their own, leading us in a big circle up the steep hillside and through the forest. As Giri indicated, it is all such familiar ground that everyone, both people and cows, knows where they are and where they are going. Taking a nap at midday is a measure of one's sense of confidence from knowing the place and the cows that allows one to feel secure enough to drop off to sleep.[7]

The bamboo collecting and basket making were a surprise as well, although most forest peoples rarely do just one thing when going out for the day. Many of the Soliga spend much of their time making baskets these days instead of grazing cattle. This raises a critical question in regard to the path taken that day: to what extent was this route a response to the presence of bamboo resources for basket making rather than a response to the presence of fodder grasses and the lack of *Lantana*?

By far the most frequently mentioned impact of *Lantana* is the reduction of forest grasses, in kind and quantity, and particularly the best fodder grasses (for example, *Naale hulu*), and so it may be that this knowledge was directly responsible for Giri's choice of route. He seemed to suggest as much when he said the route was chosen because grasses were more abundant here where there is less *Lantana* and more trees. Given the importance of basketry to this family, a survey of bamboo resources would be revealing. Giri's choice, however, presented real risks to the cattle and to us in the form of steep slopes composed of large, smooth, slippery rocks. We commonly heard other people argue that *Lantana* forces them to take increased risks with cattle to get to unimpacted areas, but Giri did not seem to raise this as a reason for his chosen route. We found only a few cases in which people had actually witnessed their cattle falling and being injured or killed as a result of grazing on steep slopes to avoid *Lantana;* so perhaps its dangers are exaggerated, which, in the interest of reducing risks, may not be such a bad thing.

Most farmers complained that they spend more time in the forest now and have to go further to find enough fodder to satisfy their hungry animals. Our trip was quite abbreviated, however, and took us not very far from home at all, although perhaps the route was altered because of us. On the other hand, the cattle did not seem to be lagging behind to eat more either and were in fact far ahead of us on the trail home. Measurement of the fullness of cattle, such as simple body weight as well as other indicators of overall health, would seem to be necessary to judge how successful each route chosen was in terms of supplying enough food.[8] Indeed, it would be useful to test the local belief that less grass in the forest owing to *Lantana* leads to weakened cows and thus greater vulnerability to disease, attack by wildlife, or injury.

Another commonly mentioned consequence of *Lantana* is the obstruction of movement, by all animals, as well as an opaqueness that reduces visibility and hence increases fear of unexpected and dangerous encounters with wildlife. While there was old evidence of elephant everywhere, and a week later we were to see a family of elephants in this very area, there was no indication that such a threat played a role in Giri's decision making on the day we went out. Interestingly, Puttamadappa claimed that the cows would smell the elephants and other danger and would avoid such areas by themselves, which again illustrates how the cows inform and mediate human–environment relationships.

Finally, Giri's own explanations for decreasing cattle numbers, implying a value change and cultural shift among the young, is supported by other interviews and household surveys that suggest cattle numbers are decreasing in favor of wage labor, agriculture, nontimber forest products, cash crops (for Lingayat), and basket making (for some Soliga), *not* because of increased mortality resulting from malnutrition and disease (Kent and Dorward 2012; Uma Shaanker et al. 2005). It may be that for those households that do keep cattle (and do not send them away to a cowshed), the forests can now provide enough grasses to support them. A survey of grasses in the forest would be necessary to test this and the other hypotheses about the relationships between changing grazing patterns; household livelihood strategies, including basketry; *Lantana;* and forest grasses.

Through this case, then, I have been able to identify a series of events that might be described as responses to *Lantana* (in other words, as direct or indirect consequences), from the choice of where to turn off the trail to how long we stayed at various locations, and even perhaps the choice by Giri to collect bamboo and make baskets. It is possible that many of the events of the day were driven by non-*Lantana* needs and constraints. As noted above, maybe the cause of the route chosen was primarily a need to make baskets rather than graze cattle, and maybe *Lantana* was an influential factor here too.

It would be interesting to construct a history of events for each of the purported explanatory hypotheses, perhaps starting with "leaving the main trail to graze cattle" if we were examining the grazing choices, or where "Giri starts collecting bamboo" if we are examining the basketry hypothesis. Tracing chains of cause and effect back in time might take us back to events that morning before we left, such as a conversation among family members, the visit of a basket retailer, or the grazing of cattle by Giri's father the day before. It is possible that Giri was able to satisfy two livelihood needs at one go via the path he chose and, furthermore, that this was possible only because this particular path took him to areas that had been least impacted by *Lantana* and thus provided both kinds of required forest products.

A unique episode of grazing is not going to answer many of my questions about response events and their causes, but even a single trip points to meaningful research paths to increase our understanding of what a cattle herder does during a particular day and whether *Lantana* or a changing culture and economy or perhaps a changing climate is playing any role as a cause of that unique sequence of decisions. What, then, are the lessons for studying the impacts of climate change on people? Responses to change are not easy to identify, nor is it easy to link any putative response to a specific cause such as biodiversity or climate change. People may have ready explanations for what they are doing, and these may take in and integrate global discourses of climate change, but they may also be imprecise or simply generalizations based on limited exposure and experience. This is not surprising given the recent onset of anthropogenic climate change and its slow, creeping, sometimes imperceptible impacts. Throughout this chapter I have emphasized the embeddedness of people in a diverse and complex context, as both Ingold and Vayda and Walters imply, which not only contributes to the uniqueness of everyday activities but also accounts for a certain lack of self-awareness among people with respect to their decision making.

For these reasons, studies of adaptation to climate change need to pay much closer attention to what people are actually doing when they make everyday decisions than to what they say they do. We need to explore the actions and reaction of hunters, fishermen, and farmers, who are still very much dependent on biodiversity. These people are often not only overlooked by governments and international adaptation policies—and thus the potential of their knowledge and innovations remains largely unknown—but also likely to suffer from generalized institutional responses to climate change impacts. This perspective is echoed in a recent review (Agrawal et al. 2012: 329): "Studies of adaptation need to be especially attentive to scale, equity, and ethical issues because, despite the global character of climate change, its consequences are produced, experienced and responded to at the local level and disproportionately by those with the least capacity to adjust."

If necessity is the mother of invention, and increasing uncertainty is going to create an as-yet-unknowable set of new needs, we can be sure that the unfolding future is going to be more varied and filled with more creativity, innovation, and ingenuity than we could ever imagine today. In order to capture and understand this potential dynamism in human adaptation to climate change, my suggestion here is to remind us of the value of a more traditional anthropology—one that uses fine-grained ethnographic approaches to follow, observe, and explain what people actually do as they live their lives. In this way we can identify new impacts and new responses; we can tease out the way ecological, social, economic, cultural, and political factors interact, sometimes in very complex and nuanced ways, to determine life's unique everyday activities; and we can, perhaps, better support those "experts of contingency" as they move into the future.

NOTES

I thank both Kumar Chinmayee for his assistance in the field and the staff at ATREE, Bangalore, for support: Ramesh Kannan, Harisha RP, Narayan, and Madesha. I am grateful to Andrew Mathews, Michael Dove, and Jessica Barnes for their comments on earlier drafts and for comments presented orally by participants at the workshop Climate Change and Anthropology, held at Yale in April 2012. This research was funded by an Ecosystem Services for Poverty Alleviation (ESPA) Project Framework Grant on Human Adaptation to Biodiversity Change (NE/I004149/1). ESPA is a program funded by the United Kingdom's Department for International Development (DFID), Economic and Social Research Council (ESRC) and Natural Environment Research Council (NERC). The views and any errors or omissions are the author's sole responsibility and do not represent those of any other parties.

1. The project and its researchers recognize that no community is truly isolated from the influence of the state. Even without corporeal presence, its policies have indirect effects on thought and action at all levels of society. That said, we are interested in places where state programs to mitigate climate change and its impacts are absent. Autonomous adaptation in these cases may be influenced by the presence and knowledge of the state in many other aspects of daily life.
2. The fifteen-month project (2011–12) joined partners from anthropology, economics, and ecology at Kent, SOAS, and Oxford with ATREE in India and Rhodes University in South Africa. Publications and works in progress can be found on the HABC website: HABC.eu.
3. The interactions between invasive and alien species and climate change, two major drivers of biodiversity change, have also been identified as critical areas for future research (Masters and Norgrove 2010).
4. The field site village is not participating in one of ATREE's award-winning programs to develop small-scale furniture and handicraft enterprises using *Lantana* (Kannan et al. 2008).
5. As a Scheduled Tribe, all Soliga households are classed as being below the poverty line and thus received up to 25 kg rice, 8 kg wheat, and 3 kg sugar at token prices. A high proportion of Lingayat households are also assigned to this category.

6. By comparison, Soliga residents in the nearby Biligiri Rangaswamy Temple Hills claim that the passing of the Wildlife Protection Act of 1972, which forbade the burning of forest undergrowth for swidden farming and the promotion of useful forest vegetation, has facilitated an extensive *Lantana* invasion (Desai and Bhaya 2009).
7. Several Lingayat farmers portrayed "bad grazers" as those who do not pay attention, who sleep, socialize, or do other things when they should be minding their cattle, protecting them from wildlife attacks, and preventing them from wandering off and injuring themselves. No doubt such statements were jabs aimed at the Soliga, whom they disparage as incompetent farmers and grazers. In our limited experience, both Soliga and Lingayat herders were equally attentive and sometimes inattentive during a day of grazing.
8. In the evening many families have to supplement cattle's diets with fodder collected in and around their fields.

REFERENCES CITED

Agrawal, A., M. C. Lemos, B. Orlove, and J. Ribot. 2012. "Cool Heads for a Hot World—Social Sciences Under a Changing Sky." *Global Environmental Change* 22, no. 2: 329–31.

Aravind, N.A., D. Rao, G. Vanaraj, R. Uma Shaanker, K. N. Ganeshaiah, and J. G. Poulsen. 2006. "Impact of *Lantana camara* Invasion on Plant Communities in the Male Mahadeshwara Hills Forest Reserve, South India." In *Invasive Alien Species and Biodiversity in India*, ed. L. C Rai and J. P. Gaur, 1–13. Varansi: Botany Department, Banares Hindu University.

Aravind, N. A., D. Rao, K. N. Ganeshaiah, R. Uma Shaanker, and J. G. Poulsen. 2010. "Impact of the Invasive Plant *Lantana camara* on Bird Assemblages at Malé Mahadeshwara Reserve Forest, South India." *Tropical Ecology* 51, no. 2S: 325–38.

Bhagwat, S. A., E. Breman, T. Thekaekara, T. F. Thornton, and K. J. Willis. 2012. "A Battle Lost? Report on Two Centuries of Invasion and Management of *Lantana camara L.* in Australia, India and South Africa." *PLoS ONE* 7, no. 3: e32407.

Crate, S., and M. Nuttall, eds. 2009. *Anthropology and Climate Change: From Encounters to Actions.* Walnut Creek, Calif.: Left Coast Press.

Cronk, Q. C. B., and J. L. Fuller. 1995. "Lantana camara." In *Plant Invaders*, 82–87. London: Chapman and Hall.

Danielsen, F., M. Skutsch, N. D. Burgess, P. M. Jensen, H. Andrianandrasana, B. Karky, R. Lewis, J. C. Lovett, J. Massao,Y. Ngaga, P. Phartiyal, M. K. Poulsen, S. P. Singh, S. Solis, M. Sørensen, A. Tewari, R. Young, and E. Zahabu. 2011. "At the Heart of REDD+: A Role for Local People in Monitoring Forests?" *Conservation Letters* 4: 158–67.

Desai, Meetu, and Sreetama Gupta Bhaya. 2009. "Community-Centred Integrated Biodiversity Management of Forested Landscapes in the Western Ghats. Final report 2005–2008. Submitted to The Sir Dorabji Tata Trust." Bangalore: ATREE.

Dooley, K., T. Griffiths, F. Martone, and S. Ozinga. 2011. *Smoke and Mirrors: A Critical Assessment of the Forest Carbon Partnership Facility.* FERN and Forest Peoples Programme, February.

Dove, M. R. 2007. "Volcanic Eruption as Metaphor of Social Integration: A Political Ecological Study of Mount Merapi, Central Java." In *Environment, Development and Change in Rural Asia-Pacific: Between Local and Global*, ed. J. Connell, and E. Waddell, 16–37. London: Routledge.

Garay-Barayazarra, G., and R. K. Puri. 2011. "Smelling the Monsoon: Senses and Traditional Weather Forecasting Knowledge Among the Kenyah Badeng Farmers of Sarawak, Malaysia." Special issue: *Indian Journal of Traditional Knowledge.* "Traditional Knowledge and Climate Change" 10, no. 1: 21–30.

Howard, P. 2009. "Human Adaptation to Biodiversity Change: Facing the Challenges of Global Governance without Science?" Paper presented to the Amsterdam Conference on the Human Dimensions of Global Environmental Change, Amsterdam, Netherlands, December 2–4.

———. 2012. "Human Adaptation to Biodiversity Change: Building and Testing Concepts, Methods, and Tools for Understanding and Supporting Autonomous Adaptation." ESPA Project Highlight, *ESPA Newsletter*, March. Accessed April 16, 2012. http://www.espa.ac.uk/newsletter/march-2012#Highlight.

Huffington Post. 2012. "An Early Bloom for Washington." Posted April 5. Accessed April 13, 2012. http://www.huffingtonpost.com/vicki-arroyo/cherry-blossom-festival--2012_b_1406246.html.

Hulme, M. 2012. "'Telling a Different Tale': Literary, Historical and Meteorological Readings of a Norfolk Heatwave." Special issue: "Cultural Spaces of Climate," ed. Georgina Endfield. *Climate Change* 112: 1–17.

Ingold, T. 2000. *Perception of the Environment: Essays on Livelihood, Dwelling and Skill.* London: Routledge.

———. 2007. *Lines: A Brief History.* London: Routledge.

———. 2011. *Being Alive: Essays on Movement, Knowledge and Description.* London: Routledge.

Kannan, R., N. A. Aravind, G. Joseph, K. N. Ganeshaiah, and R. Uma Shaanker. 2008. "Lantana Craft: A Weed for a Need." *Biotech News* 3: 9–11.

Kannan R., C. M. Shackleton, and R. Uma Shaankar. 2013. "Reconstructing the History of Introduction and Spread of the Invasive Species Lantana at Three Spatial Scales in India." *Biological Invasions* 15, no. 6: 1287–1302.

Kent, R., and A. Dorward. 2012. "Biodiversity Change and Livelihood Responses: Ecosystem Asset Functions in Southern India." Working paper, Centre for Development, Environment and Policy, SOAS, University of London.

Lipset, D. 2011. "The Tides: Masculinity and Climate Change in Coastal Papua New Guinea." *JRAI* 17, no. 1: 20–43.

Maness, H., P. J. Kushner, and I. Fung. 2012. "Summertime Climate Response to Mountain Pine Beetle Disturbance in British Columbia." *Nature Geoscience* 6, no. 1: 65–70.

Masters, G., and L. Norgrove. 2010. "Climate Change and Invasive Alien Species." CABI Working Paper 1.

Morab, S. G. 2003. "Soliga/ Soligaru." In *People of India: Karnataka*, ed. B. G. Halbar, S. G. Morab, Suresh Patil, and Ramji Gupta, 1381–89. New Delhi: Affiliated East-West Press.

Morgan, B. 2011. "Don't Ignore REDD's Impacts on Communities!" *Inside Indonesia*, July 25.

Morton, J. F. 2007. "The Impact of Climate Change on Smallholder and Subsistence Agriculture." *Proceedings of the National Academy of Sciences* 104, no. 50: 19680–685.

O'Brien, K. L. 2009. "Do Values Subjectively Define the Limits to Climate Change Adaptation?" In *Adapting to Climate Change: Thresholds, Values, Governance*, ed. W. Neil Adger, Irene Lorenzoni, and Karen L. O'Brien, 164–80. Cambridge: Cambridge University Press.

Orlove, B., E. Wiengandt, and B. H. Luckman, eds. 2008. *Darkening Peaks: Glacier Retreat, Science and Society*. Berkeley: University of California Press.

Parmesan, C., and G. Yohe. 2003. "A Globally Coherent Fingerprint of Climate Change Impacts Across Natural Systems." *Nature* 421: 37–42.

Puri, R. K. 2005. "Deadly Dances in the Bornean Rainforest: Hunting Knowledge of the Penan Benalui." Leiden: KITLV Press.

———. 2007. "Responses to Medium-Term Stability in Climate: El Niño, Droughts and Coping Mechanisms of Foragers and Farmers in Borneo." In *Modern Crises and Traditional Strategies: Local Ecological Knowledge and Island Southeast Asia*, ed. Roy Ellen, 46–83. Oxford: Berghahn.

———. 2012. "Looking for Soliga Responses to Lantana Invasion." Paper presented at International Congress of Ethnobiology, May 22. Montpellier, France. Working Paper No. 1. CBCD, SAC, Kent.

Rayner, S. 2003. "Domesticating Nature: Commentary on the Anthropological Study of Weather and Climate Discourse." In *Weather, Climate, Culture*, ed. S. Strauss and B. Orlove, 277–90. New York: Berg.

Roncoli, C., T. Crane, and B. Orlove. 2009. "Fielding Climate Change in Cultural Anthropology." In *Anthropology and Climate Change: From Encounters to Actions*, ed. S. Crate and M. Nuttall, 87–115. Walnut Creek, Calif.: Left Coast Press.

Salick J., and N. Ross. 2009. "Traditional Peoples and Climate Change: Introduction." *Global Environmental Change* 19: 137–39.

Sharma, G. P., A. S. Raghubanshi, and J. S. Singh. 2005. "Lantana Invasion: An Overview." *Weed Biology and Management* 5: 157–65.

Strauss, S., and B. Orlove, eds. 2003. *Weather, Climate, Culture*. New York: Berg.

Sundaram, B., and A. J. Hiremath. 2012. "*Lantana camara* Invasion in a Heterogeneous Landscape: Patterns of Spread and Correlation with Changes in Native Vegetation." *Biological Invasions* 14, no. 6: 1127–41.

Thornton, T., and N. Manafsi. 2010. "Adaptation—Genuine and Spurious: Demystifying Adaptation Processes in Relation to Climate Change." *Environment and Society: Advances in Research* 1: 132–55.

Uma Shaanker, R., K. N. Ganeshaiah, S. Krishnan, R. Ramya, C. Meera, N. A. Aravind, G. Vanaraj, J. Ramachandra, R. Gauthier, J. Ghazoul, N. Poole, and B. V. Chinnappa Reddy. 2005. "Livelihood Gains and Ecological Costs of Non-Timber Forest Product Dependence: Assessing the Role of Dependence, Ecological Knowledge, and Market Structures in Three Contrasting Human and Ecological Settings in South India." *Environmental Conservation* 31: 242–53.

Uma Shaanker, R., G. Joseph, N. A. Aravind, R. Kannan, and K. N. Ganeshaiah. 2010. "Invasive Plants in Tropical Human-Dominated Landscapes: Need for an Inclusive Management Strategy." In *Bioinvasions and Globalization: Ecology, Economics, Management, and Policy*, ed. C. Perrings, H. A. Mooney, and M. H. Williamson, 202–19. New York: Oxford University Press.

Vayda, A. P., and B. B. Walters, eds. 2011. *Causal Explanation for Social Scientists: A Reader*. Lanham, Md.: AltaMira Press.

Walters, B. B. 2012. "An Event-Based Methodology for Climate Change and Human-Environment Research." *Geografisk Tidsskrift: Danish Journal of Geography* 112, no. 2: 190–98.

Watts, M. 2011. "On Confluences and Divergences." *Dialogues in Human Geography* 1: 84–89.

Wilk, R. 2009. "Consuming Ourselves to Death: The Anthropology of Consumer Culture and Climate Change." In *Anthropology and Climate Change: From Encounters to Actions*, ed. S. Crate and M. Nuttall, 265–76. Walnut Creek, Calif.: Left Coast Press.

Zhou, X. N., G. J. Yang, K. Yang, X. H. Wang, Q. B. Hong, L. P. Sun, J. B. Malone, T. K. Kristensen, N. R. Bergquist, and J. Utzinger. 2008. "Potential Impact of Climate Change on Schistosomiasis Transmission in China." *American Journal of Tropical Medicine and Hygiene* 78, no. 2: 188–94.

Zuray, S., R. Kocan, and P. Hershberger. 2012. "Synchronous Cycling of Ichthyophoniasis with Chinook Salmon Density Revealed During the Annual Yukon River Spawning Migration." *Transactions of the American Fisheries Society* 141, no. 3: 615–23.

Chapter 11 Climate Shock and Awe: Can There Be an "Ethno-Science" of Deep-Time Mande Palaeoclimate Memory?

Roderick J. McIntosh

In the classic Malian film of 1995 *Guimba, un tyrant, une époque*, the director Cheick Oumar Sissoko delves into Mande deep-time conceptions of memory, at once of topography and of climate and of social authority. The Mande, peoples occupying roughly two-thirds of West Africa, speak related languages and share a deeply embedded set of concepts about moral obligations to community and about the (occult) power of landscape. In Sissoko's powerful film, Guimba is the tyrant, the king that Mande society is in important ways structured to resist or, at least, to abbreviate his reign of terror. Guimba's cruelty and social transgression will eventually lead to his humiliation and death. But until that happens, with his dark transformation of the Mande occult he defeats one after another of the powerful groups of social actors sent to bring him down: traditional nobles, vital if callow youth, horse cavalry, strangers bringing magic from afar, and even, for a while, hunters representing the most powerful, most secret power guild. He is master of the occult, transformed by social taboos, butchering his family and disdaining kinship obligations so fundamental to Mande society.

Key to this chapter, Guimba's evil is "harvested" from a place. The film is set in Jenne, in the Middle Niger of Mali, which all Mande recognize, to this day, as a seat of immense occult presence. Guimba's genius is to pervert the social power harvested from this place, perverting what should be the Mande Hero's role to bend natural elements to do good for the community, instead terrorizing it. In this sense he is an essential Mande Hero, wielding control over weather/climate as just one of the natural elements at his command. His evil is manifest in causing a solar eclipse and foiling the assault by the hunters by creating terrifying thunder, lightning, and torrential rainstorms and by conjuring at will "les sorcières" (dust devils, sure to empty my excavation units of workmen when they form at my site of Jenne-jeno). Sissoko uses comedic drama to explore the downfall of a twisted Hero, who is a Mande Weather Man of the worst kind: his perversion of power puts whole communities at risk. But the film is also about dangerous places where the morally neutral occult rests in great quantities, to be harvested during episodes recorded in social memory. Thus, throughout the lands of the Mande, changes of climate—from the shock of dust devils to the awe brought on by decades-long droughts—are powerful metaphors woven together into the deep-time memory that one can regard as a Mande "ethno-science" of climate causation.

One of the most intractable problems in archaeology is that of deep-time memory. Attempts by prehistorians to run with the proposition that peoples can package, archive, and transmit a historical account of past episodes of stress and surprise in their biophysical or social environments have not been entirely successful. The assessments of how peoples implemented successful or failed responses to such episodes in the often remote past have not been all that well received either (McIntosh et al. 2000). Yet a fundamental thesis of the ethnohistory of many peoples, including the Mande of West Africa, is that a deep-time memory is absolute reality. Many Mande, including several of my collaborators and codirectors from over thirty-five years of excavations in Mali, are affronted and insulted by the assertions of Western historians, ethnographers, and prehistorians that their deepest understandings of their past are simply contemporary contaminations of stories distilled from a formless mass of factoids, some based in history, some imagined, but at root stories about the past fabricated to serve power relations of the present. The central proposition of this chapter is that one can consider the explanations of causation in the natural world, which peoples craft when they delve into their deep-time social memory, as constituting the soul of ethno-science.

The Mande of arid West Africa, including the Sahel, live with precipitation and fluvial regimes that are arguably the least predictable in the world. Western scientific understanding of the palaeoclimatic patterning and forcing mechanisms has improved greatly. Can one imagine an integrated historical ethno-science linking strong Mande beliefs and values to these deep-time lessons about the world in which their ancestors lived? Put another way: Can archaeologists develop research programs—a cumulative design of ever more sophisticated hypotheses to be tested in the field—derived from an ethnohistorically informed understanding of how Mande peoples, today and in the distant past, comprehend the vitalizing forces of climate change? and, most important, how that apprehension fashions the actions those peoples take and took in the face of climate shock and awe? It is those actions—or, rather, the imprint of those actions on the material world—that archaeologists can dig up.

All archaeologists can do or have ever done is to analyze the patterns left in the material world by the actions of past peoples acting upon the world as they perceived it. Peoples have always acted in the face of climate and environmental change. And they will act to mitigate its effects, to put an end to climate surprises in accordance with their socially constructed perception of its causes. If such actions have produced archaeologically discernible patterns, some may appear to be incongruent and make little or no sense and be inexplicable in terms of responses to climate change that are based on Western science or Western experience. One solution would be to simply say such patterns are anomalous and leave it at that. Another might be to reverse-engineer a set of field-testable hypotheses from ethnohistorical explanations of how the world and its physics of climate work. Such an ethnoscience would rely on scientific canons of hypothesis testing to try to obtain a modicum of empirical verification of recordings in the material world of what were essentially actions based in the unrecorded perceptions and intentions of people long dead.

The Mande peoples have such a deep belief in the continuity of the moral underpinnings of their actions that it affords a good opportunity to experiment with the idea of deep-time memory. Indeed, the Mande term *Kuma Koro*—roughly, the ancient speech of the ancestors motivating the moral actions of the living—is considered by the Mande to be the defining criterion of what it means to be Mande (McIntosh 2000). Kuma Koro is not in any way a stagnant or essentialist concept: change is tradition; adaptation to a changing world is ever present. The living simply need signposts to right actions along their way as they navigate their richly textured physical, symbolic landscape.

Such deep-time memory is all about sustainability. Here I use the rather minimalist definition of sustainability as the ability of a system or society to maintain a desired way of life (McIntosh 2006b). To maintain sustainability one must be resilient: able to respond to, absorb, and creatively transform the shocks and surprises supplied by one's environment. Perhaps ironically, the Mande would be the first to say that the key to sustainability in their surprise-laden environment is change, morally directed change, perhaps, but a resilience underscored by a flexibility born of deep history. In this regard, Kuma Koro works as a guide to flexible, sustainable behavior in the face of drastic climate surprise, based on a socially accepted understanding of who has the authority to make decisions about how to act. Those decisions may be made from a palette of successful actions drawn from the past, but the genius of the Mande is all about who has the authority to persuade the rest of society to act.

Two issues are critical to any analysis of how a local ethno-science of climate might work: the nature of climate change, or the shock and awe of this chapter's title in Mandelands (for more, see McIntosh 2006a); and the derivation from deep-time memory of the moral authority of those who can persuade, the Mande Weather Machines, or the heroes and heroines of past crises whose authority carries through to today's Weather Machines (for more, see McIntosh 2000). What I will concentrate on is the typology of features in the symbolic landscape that are differentially and preferentially harvested by those heroes and heroines to add to their knowledge and also to augment their moral authority.

The larger issue for those who think about contemporary discussions of climate change is that the Mande ethno-science of climate compels us to ask our own questions about who is authorized by any society, contemporary American or prehistoric, to make decisions about how to act in the face of climate change. When captains of industry make pronouncements about climate change causation, by whose authority are those pronouncements acted on or become the basis for inaction?

THE NATURE OF PAST CLIMATE CHANGE: SHOCK AND AWE

The shock and awe of Sahelian climate are expressed along many dimensions and many scales. To begin with the shortest scales, the interannual and interdecadal: Jean Koechlin (1997: 12–18) has called the arid West African climate among the world's most variable and unpredictable. Shukla provides the classic description of the instrumental record: "In the recorded meteorological data for the past 100 years, there is no other region of the globe of this size [besides the Sahel] for which spatial and seasonal averaged climatic anomalies have shown such persistence" (1995: 44). Taking a longer chronological and a continental-scale perspective, Sharon Nicholson writes, "The largely semi-arid African continent has undergone extreme climatic changes which are probably unmatched in their magnitude and spatial extent" (1994: 121). Given this relentless saga of unpredictable weather and climate, the Mande peoples have developed a history of progressive strategies of flexibility in the face of long-term climate rhythms that can incontestably be described as being at the edge of chaos.

The Mande of West Africa occupy roughly one-third of the area of central and northern West Africa, from the semiarid tropical mixed wooded and grass savanna (roughly to the 1000mm rainfall isohyet, with short, intense rainy seasons) through the Sahel–desert interface (roughly 150mm). Ancient settlements we call proto-Mande dating to the last few millennia BC can be found deep in the Sahara in formerly lush, semiarid riverine or lacustrine situations. Lacking the dendroclimatological record of the American Southwest, we must be cautious when extrapolating back in time from the superb West African instrumental data (one of the world's tightest meteorological webs dating back to the 1880s). Working at the interdecadal scale, Nicholson has identified at least six repeating climate modes, or "anomaly types," of spatially distinct precipitation patterns (Nicholson 1986, 1994, 1995). These repeat but do not recycle in any discernible order, adding to the unpredictability for those on the ground. There is some hint that each abrupt mode shift is announced by a short, destructive period of high temporal variability of rainfall. In the absence of the high-resolution lake-core sequences of many temperate countries, the available data suggest abrupt larger-scale mode shifts of climate throughout the Holocene—precipitous shifts that are linked to changes in the Milankovich cycle, solar radiation, and more distant Holocene changes in the earth's atmospheric–oceanic circulation system. From roughly 3500 BC, the Sahara dries out (nonlinearly), peoples migrate south, then north again, and finally further to the south, especially to the floodplain refugia of the Senegal and Niger river plains and the great Lake Chad basin, all the while having to respond to great climate unpredictabilities at interdecadal and interannual timescales. Shock and awe.

I believe it is with a catastrophic dry incursion at circa 4,200–4,000 years ago that fundamental ancestral Mande attitudes to climatic stress and surprise have their birth (see McIntosh 2005: fig. 2.10). In the physical world, from this time until roughly the mid-first millennium BC, one would have witnessed rapidly changing contours of the oscillating lake and playa shores, rainfall isohyets, and vegetation zones. The counterpart in the human world must have included mass migrations and massive new accommodations to change in the biophysical sphere, including domestication, adaption of new subsistence practices and settlement patterns, and, ultimately, specialization. Further to the east, the Lake Chad basin is quite arid from 4,000 years ago until circa 3500 BP, when the lake expands in its last transgression, declining again by almost half at circa 2800 BP in its "terminal lake stage." When the dry episode of circa 4000 years ago ends, conditions return to those approximating the last pluvial. But not for long. Further Sahelian drought crises are documented between 3800 and 3600 and again at 3300 and 3000 years ago (Breunig and Neumann 2002: 125; Hassan 2002: 323, 331).

Then conditions steadily deteriorate, and in a highly oscillatory fashion, placing populations under extreme pressure to migrate or to find new solutions to food security if they wished to maintain some semblance of an older, familiar way of life. After a significant dry period around 3000 BP, sedentary, nonpastoral occupation effectively ceases in the Sahara above circa 18° N, even with a notable wet pulse at 2800 BP. There was still a sudanic (lion, elephant, etc.) fauna in many Sahelian locations and even in southern Saharan areas during the penultimate millennium BC (and even large fish, crocodile, and hippo in the better-situated lakes and seasonal streams). But evidence for these fauna evaporates like the water in the once-permanent lakes, which now become seasonally expanding and contracting playas, soon to be sad, salty *sebkhas.* Dunes remobilize. Rainfall at Tichitt in Mauritania was perhaps at 200 percent of present levels just after 4000 BP, but by 3000–2600 BP it had declined to 125 percent or less (Munson 1981).

This was the period of the first serious occupation of the Middle Niger and of the Middle Senegal to the west (McIntosh 1999). All the climate oscillations of the circa 4000 to 2300 BP period would have made these millennia prime for pulse migrations of Late Stone Age peoples out of the desiccating Sahara and drying playas, streams, and grasslands of the Sahel into the permanent refuge of the Middle Niger floodplain and to sudanic regions further to the south. In now-forested Ghana, Lake Bosumtwi is in deep regression from 2500–2000 BP. Vernet (2006) speaks of an abrupt Sahelian drying and even retreat of the equatorial forest beginning at 2600, with a finality to the Sahara's fate. As Vernet laments, "Un seuil biologique est franchi" (a biological threshold has been crossed).

By around 300 BC and until circa AD 300 West Africa shared with many other tropical regions (Southeast Asia and the Amazon Basin, among others) a stable dry phase. Precipitation was somewhat or moderately below present levels for a half millennium or longer. Perhaps its very stability aided the massive influx of peoples into the still-well watered southern regions. In the Middle Niger, a purely indigenous, Mande urbanism takes root. This protracted dry period is followed by a relatively stable period of increasing precipitation (to perhaps +125 to 150 percent of today) to AD 700, when rainfall maintains a stable plateau until circa AD 1100. Lake Bosumtwi comes out of regression at circa AD 200, and within a century the northern basin of Lake Chad, the Bahr el Ghazal, is fed by discharge from the northern highlands (highstand at the seventh century). By AD 500 some Saharan highlands to the west (the Tagant) even have a vestigial Sahelian reoccupation up to 20° N. This was a time of relative stability, that is, relative to the dynamic instability of the phase to follow. However, surely more droughts will be recognized as our resolution at the 10^4 timescale improves.

It seems to have been the case in the Sahel that, in places as early as AD 900 but in others perhaps as late as 1100, there was a demonstrable "catastrophic discontinuity" between this and the following unstable centuries. In the Middle Niger, to date, there is little to warrant subdividing this AD 1000 to 1860 phase, although I suspect this will be remedied with improving resolution. It is worth recalling that Lake Chad is in transgression (minor, to be sure) for perhaps two centuries after circa AD 1000. Recall also that this unstable phase corresponds to the better-understood 10^4 timescale climate oscillations of Europe—the Medieval Warm period of AD 1290–1522 and the Little Ice Age of AD 1550–1860. The middle and upper latitude record for the Medieval Warm and Little Ice Age have come under detailed scrutiny of late (Dunbar 2000; Ruddiman 2001: 356–81). Although the tropical record is comparatively spotty, there is now an extraordinary subdecadal resolution of the monsoon over the Cariaco Basin off the Venezuelan coast, identified in high-resolution sea cores, with strong teleconnection associations with the Sahelian monsoon (Haug et al. 2001; see McIntosh 2005: fig. 2.9b). While it would certainly, unfortunately, be too much to ask for decade-by-decade exact correlations with the Sahel, it is useful to appreciate the abrupt-onset shifts and phases of variability in this extraordinary high-resolution record of monsoon variability from the other side of the Atlantic.

The Medieval Warm Period begins abruptly and with a short interval of high variation. Precipitation is high and stable. High amplitude variability marks the close of this and the beginning of the Little Ice Age, with particularly high amplitude characterizing the time since 1600. Indeed, Brooks (1998) reasonably insists upon early second millennium AD diminishing rainfall and shifts of ecological zones several hundred kilometers to the south throughout Mande country. Then, still in northern West Africa, Brooks speaks of higher precipitation and a disastrous spread of tsetse 200km to the north, from circa 1500 to 1630. Brooks then posits a dry, drought and famine period from 1630 to the end of the nineteenth century, with particularly severe droughts in the 1640s, 1670–80s, 1710–50s, 1770–80, and especially 1790–1840 (see also Webb 1995: 8). After circa 1650, we can hope to rely as never before on historical records of droughts, floods, famines, and pestilence for reconstructions of patterns in the climatic data. We know historically that significant oscillations of West African lake levels of a few decades in length occurred at 1680–90, 1740–60, and 1800–40, to take the example of Lake Chad declines—but did the lake oscillate as often and as deeply in the millennia before? However, it is at the next shortest timescale—of events resolvable at the century scale—that the most spectacular advances have been made. In particular, with those historical records we can begin to compare what Nicholson (1986) calls "anomaly types" tickled from the last century instrumental records with the historical sources, sketchy though they may be.

In an extensive study of historical, instrumental weather data from all over the African continent, using colonial records of really quite surprising coverage going back into the late nineteenth century, Nicholson (1986, 1994, 1995) has posited phase-like regularities in continental-wide rainfall variability. She has published six reoccurring precipitation patterns, abrupt-onset modes, or anomaly types, through which African climate has recycled repeatedly well into the nineteenth century. For example, Mode One describes the dry conditions over virtually all of West Africa that characterized the infamous Sahel Drought, beginning in 1968–69. Mode One also describes drought conditions in the last decade of the nineteenth century, lasting until 1917, and also similar conditions from circa 1828–39. Mode Three, on the other hand, describes the unusually wet conditions over the Sahel and southern Sahara of the 1950s and into the early 1960s. Disastrously, these were the anomalously wet conditions that encouraged an expansion of herds, leading to severe overgrazing, livestock losses, and landscape damage when the Sahel Drought hit at the end of that decade. As our historical and archaeological deep-time climate data become higher in resolution, we will test just how far back in time these anomaly types, or precipitation modes, can be extrapolated. But for the moment it is a reasonable working hypothesis that, at the centennial timescale, Mande climate leapt in a virtually random, unpredictable manner from one mode (one statistical range of variability) to another.

There are two takeaway lessons from Nicholson's six modes. First, in terms of human response, farmers, herders, and fisherfolk, each in their own way, have had to deal not with smooth and gradual changes in the on-the-ground realities of climate variability but with surprise. Abrupt-onset . . . shock and awe! Nicholson analyzes the leaps from one type of anomaly to another in terms of abrupt-phase transitions as Mande climate was forced across thresholds. At centennial and interdecadal timescales, change is not just unpredictable, it is radical as well. Second, if one is to be confronted during one's lifetime with three or more radical departures from "normal" climate, would it not make perfect sense for one's society to have developed mechanisms to record, archive, and transmit to following generations past experience with those recycling modes? This is one of the geniuses of the Mande peoples (explored further in McIntosh 2005, esp. chapters 2 and 3). I turn now to social memory as a particular form of adaptation.

MANDE WEATHER MACHINES

Ancient Mande peoples adapted to such shock and awe in a variety of ways, for example, by migration and by adopting more arid-adapted agropastoral or full pastoral economies. Some, especially in the great floodplain of the Middle Niger, developed a highly successful and sustainable urbanism that was based, we believe, on a complementary hyper-specialization of corporate groups linked in a regionally generalized reciprocating economy. Put most simply, a good year for the swamp fishermen (Bozo) might be a mediocre year for farmers (Nono, Marka) of floating African rice (*Oryza glaberrima*) and a disaster of a year for rain-fed millet farmers (Bambara). The knowledge load required for the successful farming, fishing, and herding of local species in this highly unpredictable region is enormous, making specialization a desirable thing. But such hyper-specialization is sustainable only if social memory acts as, in effect, a contract ensuring reciprocal sharing among the various corporate groups, which, through time, came to include specialized potters, blacksmiths, leather workers, and an expanding list of craft persons (McIntosh 2005).

Thoroughly interwoven in this skein of specialists is an over-fabric of climate specialists, the Mande Weather Machines: habitually hunters, including hunters of aquatic animals, and blacksmiths, but they can come from any corporate group. Most are believed to create weather quite handily—especially torrential rains and whirlwinds—as depicted in the film *Guimba the Tyrant.* These heroes and heroines, aka Mande Men and Women Weather Machines, derive their knowledge and their authority by traveling to sanctified places or making elaborate pilgrimages (English does not suffice to convey the concepts here)—the Mande term for which is *Dali masigi*—a highly dangerous knowledge quest. The hero/heroine must leave the comfort of home, roam strange lands inhabited by friendly or, more likely, unfriendly spirits (rather, avatars of place), sometimes confronting those spirits in combat or simply rolling around in the dust of a forbidden cave. He/she comes back altered forever. She/he comes back with special knowledge/moral persuasion to help society act in the face of abrupt climate and environmental surprise.

The Mande among themselves are in endless conversation about the pinnacles, valleys, plateaus, and depressions in the landscape navigated by Mande heroes and heroines (fig. 11.1). The physical landscape of the Mandelands is remarkably flat; but the symbolic landscape is a riot of topography, one that can be augmented by the moral acts of the Mande Men/Women of Crisis or depressed by acts of evil. Thus, the Mande conceive of their landscape as a topography of peaks, locales where social order brings benefits to all; valleys, where things just fall apart; and crevices, where moral evil, past or present, constitutes an ongoing danger. The concept of this moral power is called *Nyama.* To paraphrase Pat McNaughton (1988: 16) "[The three-dimensional mapping of *Nyama* over the landscape] is a little like a network of high-tension wires strung out over the landscape." There may be endless discussion about which locale is more *nyama*-laden or more dangerous for mere mortals (or which foreign archaeologists must be forbidden from seeing, much less from striking with a shovel!). But everyone agrees on the basic workings, the "ethno-physics," of the symbolic landscape. There is a peppering of places, some open to all, some hidden and far too dangerous for everyday folk; each with its own topographical knowledge to impart to the visitor prepared and morally primed to receive that occult knowledge; each playing its unique part in the drama of heroes marching across the landscape to a place of greater power and greater esteem within society. A little conversation with a genie here, a little extra knowledge of poisons there, a roll in the dust of an

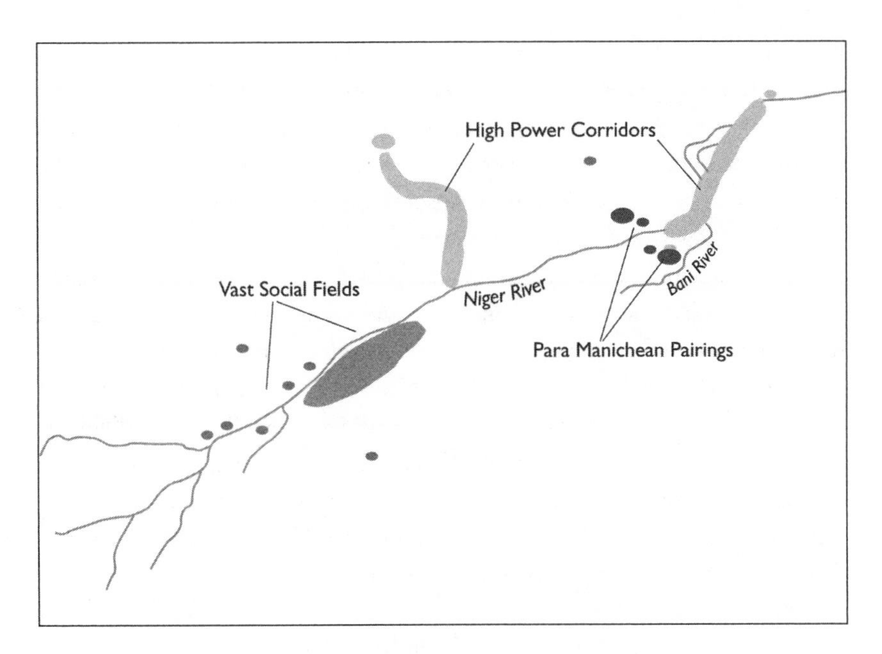

11.1: Mande sacred landscape. Bill Nelson.

incredibly dangerous spot imparting to the "dustee" greater protection from an enemy's spite. We have a dappled landscape of sacred, powerful, dangerous sites: all with power to be harvested by heroes and heroines; each differentially mentioned in the moral stories of different clans and families and in the sagas of the major dynasts of the emerging Mande polities; each evoked in the foundation epics of the great Ghana, Mali, Songhai empires—each a classic appropriation of the power of place.

The places that interest me most are the ironworking localities. One can find smelting sites with mid–first millennium AD or even early first millennium BC dates like the ones I have excavated on the Middle Niger (McIntosh 2005) or from circa 800–500 BC like the ones at Walalde on the Middle Senegal (Deme and McIntosh 2006). But can we assert with confidence that these ironworking localities were as *Nyama*-packed (as occult-power packed) as they would have been in the last century, much less in Sunjata's, the founder of the Mali empire, thirteenth century? Where is the empirical proof?

Now to the reverse engineering: what are some common characteristics of ethnohistorically *Nyama*-rich places that may alert the archaeologist to be on the lookout for other anomalies in the entire assemblage of settlement and material data from that locale? To start with the most intractable: What do we do about the Land of Do and Kri, across the Niger and south of Mali's capital, Bamako? Do and Kri—one could almost call it the Land of Lies and Disinformation. A vast field of occult power and posturing, yet the legends and memories that we have recorded are contradictory. It is almost as if the *jelli*, the keepers of oral traditions in southern Mande, had a conference a hundred years ago and agreed to weave a tissue of lies, a complete fabrication to lead outsiders in the opposite direction of the truth, so that the cosmic symbolism of the place would lie peaceful and undisturbed as a consequence. It appears that we are dealing here with a plateau of high-charged occult power. This is unlike the power locales of the Mande heartland surrounding the Mali–Guinea frontier, which I describe below. There, I think individuals or clans want to be associated with the occult potential of specific places, so ethnohistorians and archaeologists benefit from a certain amount of explicit boasting. But as for Do and Kri—there just may be places too secret, too dangerous for us ever to get an empirical handle on them.

Easier than Do and Kri and something I know we archaeologists can deal with is the Vallée du Serpent. This is a long corridor going from the left bank of the Niger all the way up to Koumbi Saleh, the putative capital of the Ghana empire in Mauritania. The traditions about the Vallée du Serpent state that a giant malevolent snake, which could very well be the Bida of Ghana lore, passes underground from Koumbi to the Niger River. No Mande in his or her right mind would live on the earth's surface above that subterranean passage. Several years back Kevin MacDonald did a survey of the Vallée du Serpent and found an inexplicable settlement pattern: lots of very old, stone age villages which he and Allsworth-Jones call "macrolithic" and date back to the mid-Holocene, maybe to 5000 BC (MacDonald and Allsworth-Jones 1994). And then little, if anything else until some historic ceramics that look just like the material we have in the Tukulor villages of the Middle Senegal. The Senegalese cleric-conqueror El-Hajd Umar, in 1862, comes through the Vallée du Serpent with his Tukulor troops—dropping pots along the way! The Vallée du Serpent is at the rainfall tier of the Sahel broadly similar to the Méma or the other northern palaeochannels (Gorgol, Kolimbiné, Fala de Molodo, or the Tilemsi), which have abundant Late Stone Age and Iron Age material on their banks. Could it be remotely possible that we could have a tradition of a malevolent snake and of an avoided landscape preserved over six or even seven thousand years?

The implied archaeological anomaly is interesting and worthy of further empirical investigation. As good scientists, we try our best to first eliminate the easiest, simplest, most parsimonious explanation, which would be some environmental difficulty with the Vallée du Serpent that is not expressed in analogous palaeochannels. However, it is difficult to imagine what those environmental difficulties might be. If, after this possibility is looked into, it can be eliminated, then we are obliged to fall back on explanations involving beliefs rooted in the occult—even if that forces us to imagine a long-standing belief about the land, a layered transformation in that belief over deep time.

Long, narrow corridors, often rivers or now-senescent palaeochannels, seem to figure often in tales about layered transformations of the landscape. Take the potential for this ethno-science of Jenne and Dia in the Middle Niger. The underlying premise of chapter 5 (300 BC) and even chapter 3 (4100 BC), the first two historical imagination chapters of my *Peoples of the Middle Niger* (1999), is that Jenne and Dia are fundamentally different from later Timbuktu, at least the Timbuktu of historians. Interesting research is being done by Yale graduate students at earlier cities found in Timbuktu's hinterland during surveys of long palaeochannels penetrating the desert, with their proximal ends at the Niger's Bend (Park 2010). No traditions have been collected (yet) about those. Perhaps because of the high concentration of *baafaro*, or water genies, in the surrounding waters and marigots, Dia and Jenne are places of incredible occult danger (applicable also to the whole of the Niger from Jenne downstream to Timbuktu). Maybe it is as simple as a higher rice potential in the surrounding backswamps. But I think it is clear from the archaeology of Jenne-jeno's (the site of ancient Jenne) lowest levels that even smiths were attracted to that highly unlikely terrain—no ore, no trees for charcoal, after all. If you accept that smiths were already occult masters by the beginning of the Iron Age, then it is likely that the occult-magnet of those localities was in place even earlier.

The clear difference between these two ancient towns and Timbuktu lies in what I now call (tongue-in-cheek) the "Para-Manichean pairing": Jenne and Gomitogo; Dia and Dia-Bozo. One represents order, a high potential for *Nyama*-grounded moral initiatives. The other—well, things just go wrong, things just fall apart. Not good and bad. Not light and darkness. Just positive initiatives and "blah," which defy description in English or French. Are these the signatures of the earliest cities in the Middle Niger? Are these the signatures of points on the landscape that had the potential to be transformed radically by the actions of smiths and other Mande Men and Women Weather Machines in the past? and by (Sufi) Islamic learned men (marabouts) today? Islam comes in, and there is another layered transform of the towns:

Marabouts who dabble in supernatural powers; amulets and potions. It is rich material for an anthropologist attuned to syncretistic social practices and the emergence of syncretistic beliefs.

Last, there are the peer polity, competing-and-interactive social fields. I have alluded above to the so-called Mande Heartland of equivalent villages, some with powerful smiths, some with other occult masters, where the entire terrain is an arena for *Dali masigi*, that is, the pilgrimage by Mande Heroes or Heroines across the landscape to harvest knowledge and occult power. Yet any power accumulation by one locale is kept in check by the countervailing power of the others. Likewise, from the south we have tales of Sundiata and the king of the Méma (Conrad 2004). We have none from the Méma itself, though. It is too sparsely populated now, and there is the likelihood that the original population of Mande Soninke has been largely displaced by environmental changes and by the present Fulani and Moors. From the oral traditions themselves we could never know whether stories about the role played by the Tunka—the king who provided refuge for Sundiata, who schooled him in the finer points of western Sudanese statecraft, and who provided him with the core corps of his army—had a core of historicity or whether they are just part of a larger fabric of invented tradition to legitimate the conquests of Sundiata.

Palaeoclimate research and archaeology give the details not only of how the climate and fluvial system changed in the Méma but also of how a long-term resilient social system held on and adapted until that final, rather abrupt avalanche or cascade of regional abandonment. With more survey and excavation, we will learn from the archaeology what people ate and how they lived and managed their crafts, and the like. We will be able to say much more about the local evolution of politics, society, economies, and demographies from an earlier late stone base.

But what only ethnohistory can give us is a sense of individual agency: the people and their emotions, their motivations and intentionalities. Without putting too fine a point on it, to what degree are our received Western understandings about human responsibilities for action to mitigate the effects of climate change, or rather justifications for inaction, yet another form of ethno-science? Warning: if we start from the premise that empirical science, the hypothesis testing of archaeologists when they put on their science hat, is at some fundamental basis incompatible with the "human drama" of the oral traditions and beliefs in a power landscape etched in symbols on the terrain, then we will never know, will we?

REFERENCES CITED

Breunig, P., and K. Neumann. 2002. "From Hunters and Gatherers to Food Producers: New Archaeological and Archaeobotanical Evidence from the West African Sahel." *In Droughts, Food and Culture: Ecological Change and Food Security in Africa's Later Prehistory*, ed. F. A. Hassan, 123–55. New York: Kluwer Academic/Plenum.

Brooks, G. E. 1998. "Climate and History in West Africa." In *Transformations in Africa: Essays on Africa's Later Past*, ed. Graham Connah, 139–49. London: Leicester University Press.

Conrad, D., ed. 2004. *Sunjata: A West African Epic of the Mande Peoples*. Cambridge: Hackett.

Deme, A., and S. K. McIntosh. 2006. "Excavations at Walaldé: New Light on the Settlement of the Middle Senegal Valley by Iron-Using People." *Journal of African Archaeology* 4, no. 2: 317–47.

Dunbar, R. B. 2000. "Climate Variability During the Holocene: An Update." In *The Way the Wind Blows: Climate, History and Human Action*, ed. R. J. McIntosh, J. A. Tainter, and S. K. McIntosh, 45–88. New York: Columbia University Press.

Hassan, Fekri. 2002. "Conclusion." In *Droughts, Food and Culture: Ecological Change and Food Security in Africa's Later Prehistory*, ed. F. Hassan, 321–33. Dordrecht: Kluwer Academic.

Haug, G. H., K. A. Hughen, D. M. Sigman, L. C. Peterson, and U. Röhl. 2001. "Southward Migration of the Intertropical Convergence Zone Through the Holocene." *Science* 293: 1304–07.

Koechlin, J. 1997. "Ecological Conditions and Degradation Factors in the Sahel." In *Societies and Nature in the Sahel*, ed. C. Raynaut, 12–36. London: Routledge.

MacDonald, K. C., and P. Allsworth-Jones. 1994. "A Reconsideration of the West African Macrolithic Conundrum: New Factory Sites and an Associated Settlement in the Vallée du Serpent, Mali." *African Archaeological Review* 12: 73–104.

McIntosh, R. J. 1998. *Peoples of the Middle Niger*. Oxford: Blackwell.

———. 2000. "Social Memory in Mande." In *The Way the Wind Blows: Climate, History and Human Action*, ed. R. J. McIntosh, J. A. Tainter, and S. K. McIntosh, 141–80. New York: Columbia University Press.

———. 2005. *Ancient Middle Niger*. Cambridge: Cambridge University Press.

———. 2006a. "Chasing *Denkejugu* Over the Mande Landscape: Making Sense of Prehistoric and Historic Climate Change." In *Climates of the Mande*. Special Segment of *Mande Studies* 6: 11–28.

———. 2006b. "Two Thousand Years of Niche Specialization and Ecological Resilience in the Middle Niger." In *Climates of the Mande*. Special Segment of *Mande Studies* 6: 59–75.

McIntosh, R. J., J. A. Tainter, and S. K. McIntosh, eds. 2000. *The Way the Wind Blows: Climate, History and Human Action*. New York: Columbia University Press.

McIntosh, S. K. 1999. "Floodplains and the Development of Complex Society: Comparative Perspectives from the West African Semi-Arid Tropics." In *Complex Polities of the Ancient Tropical World*, ed. E. A. Bacus and L. J. Lucero, *Archaeological Papers of the American Anthropological Association* 9, no. 1: 151–65.

McNaughton, P. 1988. *The Mande Blacksmiths: Knowledge, Power and Art in West Africa.* Bloomington: Indiana University Press.

Munson. P. J. 1981. "A Late Holocene (c.4500–2300 bp) Climatic Chronology for the Southwestern Sahara." *Palaeoecology of Africa* 13: 97–116.

Nicholson, S. E. 1986. "The Spatial Coherence of African Rainfall Anomalies: Interhemispheric Teleconnections." *Journal of Climate and Applied Meteorology* 25: 1365–81.

———. 1994. "Recent Rainfall Fluctuations in Africa and Their Relationship to Past Conditions Over the Continent." *The Holocene* 4: 121–31.

———. 1995. "Variability of African Rainfall on Interannual and Decadal Time Scales." In *Natural Climate Variability on Decade-to-Century Time Scales*, ed. D. Martinson et al., 32–43. Washington: National Academies Press.

Park, D. P. 2001. "Prehistoric Timbuktu and Its Hinterland." *Antiquity* 84: 1076–88.

Ruddiman, W. F. 2001. *Earth's Climate: Past and Future.* New York: W.H. Freeman.

Shukla, J. 1995. "On the Initiation and Persistence of the Sahel Drought." In *Natural Climate Variability on Decade-to-Century Time Scales*, ed. National Research Council, 44–48. Washington: National Research Council.

Sissoko, C. O., director. 1995. *Guimba, un tyrant, une époque.* Bamako: Kora Films.

Vernet, R. 2006. "Evolution du Peuplement et Glissement des Isohyetes à la Fin de la Préhistoire et au Début de l'Histoire en Afrique de l'Ouest Sahelienne." *Mande Studies* 6: 29–49.

Webb, J. 1995. *Desert Frontier: Ecological and Economic Change Along the Western Sahel, 1600–1850.* Madison: University of Wisconsin Press.

Afterword: The Many Uses of Climate Change

Mike Hulme

In their introduction to the history of science journal *Osiris* for 2011, whose theme was climate, Jim Fleming and Vlad Jankovic draw attention to the ways in which the idea of climate has been understood over time and across cultures. They suggest that the conventional definition of climate adopted by contemporary meteorologists—the statistical aggregation of thirty years of weather data—is anomalous when viewed historically. As they state, "Climate has more often been defined as what it *does* rather than what it *is*" (Fleming and Jankovic 2011: 2). Climate has been understood more commonly as a force, an agent, than as an index of change, an indicator of weather trends. Climate's agency, they suggest, has extended historically "as a force . . . informing social habits, economic welfare, health, diet, and even the total 'energy of nations'" (2).

Breaking with this deeper historical understanding of climate, nineteenth- and twentieth-century climatologists and meteorologists increasingly developed the idea of climate as index. For example, Alexander von Humboldt's deployment of isolines to index temperature through global maps first occurred in mid-nineteenth-century

Europe. And in the twentieth century powerful global observing systems expanded the range of indicators that could reveal climate: indexes of the El Niño–Southern Oscillation, global temperature, hurricane frequency, regional precipitation, and drought. Yet over the past quarter century both professional and public discourse around climate change shows evidence of a return to climate understood as agent as well as index. For example, the UN's Intergovernmental Panel on Climate Change (the IPCC) devotes an entire Working Group to the agential force of climate change (impacts) and how humans can dissipate this force (adaptation).

These changes in the way climate has been understood are evident in the subtly changing use of language. In the 1970s and 1980s, as concern about the changing composition of the atmosphere under the influence of human activities grew, the prospect of a warming of world climate was most commonly described as "*climatic* change." For example, the interdisciplinary journal established in 1977 by the prominent climatologist Stephen Schneider was called *Climatic Change*, and it still has that name today. In this terminology *climatic* is a qualifier of *change*, just as the adjectives *political* and *economic* qualify other types of change. Attention focused on changes in the weather as revealed by the new indexes of climate being developed.

But during the 1990s the term *climatic change* came to be increasingly replaced with the term *climate change* (for example, see Henderson-Sellers 2010). It is very rare now to find the former term being used in either academic or public discourse. Climate-change in effect becomes hyphenated. As a noun it becomes a cause of change, an agent with power, rather than a description of change. Today, one is more likely to find the idea of climate change validated through its impacts, the consequences of its action in the world, rather than through its usefulness as an index of changing weather. The term gets used as an adjective, as in "climate-changed weather," that is, weather which has been climate-changed, weather which has been subjected to the forces of climate change. It even becomes personified, as in, "It is in everyone's interest—left and right—that climate change has a strong, proud conservative face" (Corner 2013: 27). One is reminded of Jim Lovelock's mythical personification of Gaia: her "revenge" and her "vanishing face." This is agency writ large.

A recent illustration of this confusion between climate change as agent and as index occurred in the summer of 2013. In June 2013, after a series of unusually cold winters and springs and wet summers in the United Kingdom (the opposite of what climate model predictions based on rising greenhouse gas concentrations were suggesting), the UK Met Office convened a meeting of leading meteorologists to try to understand the reasons for the extreme seasonal anomalies. In an interview after the meeting, Stephen Belcher, the head of the Met Office Hadley Centre, remarked, "We'll particularly be looking at the way oceans and the atmosphere exchange heat, as well as how models capture that process, the influence of the stratosphere, and *which of the drivers we've looked at may be influenced by climate change*" (emphasis added).[1] In Bolcher's thinking, climate change takes on the role of a causative agent of interannual weather variation. This is in contrast to understanding climate change as an index of change in the climate system, to which interseasonal variations in weather would contribute.

In my afterword to this edited collection of anthropological essays on climate change I reflect on the roles of climate as index (climatic change) and as agent (climate change). I organize my thoughts around the ideas of causation, representation, and instrumentalism and draw on some of the ideas and arguments presented in the previous chapters.

CAUSATION

Although many people today present anthropogenic climate change as "the defining issue of our time," a challenge unique to modern *Anthropos*, Michael R. Dove, in chapter 1, makes the case for emphasizing continuity between current debates and past discourses about climate, change, and society. One reason the climatic future may *not* be beyond the grasp of historical sensibility is the enduring nature of the questions that are prompted by climate and its interactions with human societies: what causes climates to change? what phenomena are caused by climate and its changes? Humans have long sought to answer these questions, and their cultures have frequently been fashioned around the answers to them. Thus weather has been understood variously as "the domain of the gods" (Donner 2007), as the result of humanity's moral (mis)behavior or material activities, and as the random outworking of a complex, nonsentient physical system. As Dove explains, the age-old question, Why is the other different? lends itself to the seductive explanatory power of climate to explain difference and change in material and symbolic worlds.

But the range of phenomena which have been caused by climate or, as understood more recently, by climate change, extend well beyond questions of othering. Why are peatlands deteriorating? Why do herders' searches for pasture meet with failure? Why is water scarce? These are examples of questions of climatic cause and effect that are addressed in this book, although the list of outcomes for which climate change has been called upon to take responsibility is near endless. These three questions are reasonable ones, and what a number of the chapters above offer are accounts of whether and in what way climate and its changes might be held responsible for such observed outcomes.

In chapter 11, Roderick McIntosh approaches the question of climate–society relationships in prehistory. Based on his study of the Mande culture of Mali in Sahelian West Africa, McIntosh shows how traces of historical interactions between climate and society live on in the cultural memories of the living descendants; what the Mande call *Kuma Koro*, "the ancient speech of the ancestors motivating the moral actions of the living" (p. 275). McIntosh writes, "Throughout the lands of the Mande, changes of climate . . . are powerful metaphors woven together into the deep-time memory that one can regard as a Mande 'ethno-science' of climate causation" (p. 274). He makes a plea for recognition of the possibility that ethnohistory reveals people's past emotional and imaginative interactions with climate. McIntosh offers socially constructed understandings of the causes and consequences of climate change, yet recognizes that these understandings are entwined with the material traces of weather left in landscape and archaeology.

But the physical force of climate's variations always acts on material and cultural landscapes which are subject to multiple forces and processes. As Frances Moore and her colleagues write in chapter 7, "Any researcher seeking to identify climate change impacts in the context of the lived experience of people affected will early on run into the question, Why climate?" (p. 178). Precipitation, snow, ice, heat, wind, frost never act alone. Which is why the question that underpins Karina Yager's analysis in chapter 6 of community perceptions of climate change in Bolivia is such an important one: How does one "know" climate change is occurring? Yager's study of the Aymaran pastoralists and their peatlands in the Sajama National Park illustrates the entanglement of climate as both index and agent. Climate as index can be studied by using remote sensing technologies that may enumerate changes over time in the physical properties of atmosphere, biosphere, and cryosphere. But to establish climate change as the *cause* of peatland expansion or deterioration, local knowledge and the experiences of the pastoral herders are necessary. It then becomes immediately apparent that climate never acts alone. Social trends, livestock numbers, irrigation practices, and pasture management, together with changes in precipitation and glacier meltwater, offer a more complex account of change in a local setting.

When climate change is invoked as a cause of change, very often it is being used to stand in for a wider set of human concerns about their changing cultural, political, moral, and physical worlds. Climate change then becomes an index of—that is, it becomes a synecdoche for—a much broader sweep of portentous disturbances affecting the familiar and predictable lifeworlds humans create for themselves. This approach to climate change has been powerfully illustrated in the work of Peter Rudiak-Gould in the Marshall Islands of the Western Pacific in his book *Climate Change and Tradition in a Small Island State: The Rising Tide* (Rudiak-Gould 2013a). To say that "climate change has caused this outcome" requires at the very least a much richer account of the categories of change that are bound up in vernacular understandings of the term. And anthropologists have the skills and investigative practices to reveal what these might be (Barnes et al. 2013).

This ethnographic approach to understanding complex accounts of cause and effect is put to good use in chapter 10, where Rajindra K. Puri studies the responses of the Soliga tribals in Karnataka in southern India to the invasion of *Lantana camara* (the plant red sage). While the spread of this invasive species is said to have been exacerbated by changing climate, it is not clear either how these forest hunters and gatherers respond to this invasion or whether their responses should be deemed to be a consequence of climate change. As a way of navigating the complexities of climate change and its impacts, Puri focuses on the quotidian activities that form the basis of people's lives. In doing so he shows how context continually changes the way in which climate agency operates in local settings. Climate change impacts and responses take place within the holistic experience of forest life, encompassing changes in "biodiversity, water and soils, other subsistence activities such as agriculture, household livelihoods, kin and social structures, cosmologies, values, belief and ritual systems, local economies and markets, regional demographics, institutions and governance at all scales, development and conservation policies, and the discourse of climate change" (Puri, p. 250). The reductive claim that a given human action is allegedly caused by climate change is imprecise at best and irrelevant at worst.

REPRESENTATION

The chapters in this book grapple with the complexities of climate as agent, but another set of contributions to the book engages more with the challenges of indexing, or representing, anthropogenic climate change. In her book *Mediating Climate Change*, Julie Doyle makes the observation that responses to climate change will depend in large measure on the "institutionalised knowledge systems and discursive [and visual] practices through which climate change has come to be identified and made meaningful" (Doyle 2011: 4). Studying these systems and practices is another way in which anthropological study can contribute to a better understanding of the idea of climate change. It also shows why Moore et al.'s call in chapter 7 for a critical social science of climate change is so necessary.

McIntosh's study of the Mande offers a tantalizing way of opening up these systems by offering social memory and the muted voices of ancestors as an index of historical climate. But another set of representational practices is linked to the visual, and in chapter 2 Ben Orlove and his colleagues set out to question the ways in which climate change is made visible. Many of the most iconic (Western?) visual representations of anthropogenic climate change—polar bears, Pacific atolls, tropical drylands—enlist specific places that have historically and politically charged pasts. European nations in particular have long, complicated emotional and political entanglements with such places. Orlove makes the point that such visual cues entrain unpredictable and politically charged narratives into the story of climate change. At the very least, what climate change *means* to a given audience is altered through such representative practice. In a related context of studying the geopolitics of cartoon representations of climate change, Kate Manzo points out, "The issue . . . is not whether visuals can accurately represent the reality of climate change, but whether they can effectively represent the geopolitics of climate change" (Manzo 2012: 481). Orlove's work points to the fact that no representation of climate change is politically neutral.

As well as narrative and visual forms of indexing climate change, there are also the institutionalized knowledge systems to which Doyle refers. Knowledge is a form of representation in which some material or imaginative reality is given discursive shape, political legitimacy, and cultural status through technological and social processes. Whether knowledge is scientific, local, or personal, its making always involves human values and cultural framings. Knowledge can be lost as well as found, weakened as well as strengthened, made incredible as well as credible (Burke 2012). These are all characteristics of climate change knowledge which Jessica O'Reilly explores in chapter 4 in her case study of the knowledge culture of the IPCC and the representational practices of its scientists. She shows how projections of future climate are made—how future climate is indexed—and also how the necessary "guesswork" is represented and communicated by the scientists involved. Anthropologists are long familiar with studying informal, arcane, and mysterious practices, and O'Reilly suggests that this experience has great public value in being deployed to bring authoritative science under critical scrutiny.

The cultural dimensions of climate knowledge making are further illustrated in Myanna Lahsen's study of skeptical scientists in chapter 9. She shows how professional training, academic subcultures, and personal attributes can shape skeptical attitudes to climate science just as much as raw ideology. Lahsen argues that a much more nuanced typology of scientific positions on climate change is warranted than the crude binary divides that studies such as Anderegg et al. (2010) and Cook et al. (2013) have constructed, where in both cases climate scientists are divided too neatly into rigid positions with regard to whether or not "humans are causing global warming."

INSTRUMENTALISM

In my book *Why We Disagree About Climate Change* (Hulme 2009: 341), I talk about the plasticity of the idea of climate change—how it can be enrolled in support of "diverse projects that array themselves in the language of climate change: the creative arts, re-thinking intellectual property, securing energy, urban planning, poverty alleviation, gender equality, genetically-modified crops. We can also mould the idea of climate change, knowingly or not, to serve many of our psychological, ethical and spiritual needs."

A number of the contributions here illustrate the ways in which climate change becomes co-opted in support of different political projects. Making this observation about the instrumental uses of climate change is not pejorative—or at least not necessarily so. It would be strange indeed if the framing, production, and deployment of knowledge about climate change did not carry significant value commitments—as we have seen in the section above. But the observation is vitally important for gaining a wider appreciation of the range of subtle and not-so-subtle ways in which ideologies, power, and interests become attached to the idea of climate change. Failure to do so takes climate change—its causes and responses—beyond politics, and this surely is a dangerous place for it to be (see Machin 2013).

Anthropogenic climate change has provoked a vast array of policy instruments, regulations, and agreements, operating at multiple scales and connecting together diverse coalitions of actors. Perhaps none has been as complex to study as the policy instrument known as Reducing Emissions from Deforestation and

Degradation plus (REDD+). Two studies in this book, one about Mexico and one about Vietnam, examine the local politics of REDD+ and how climate change is deployed in the justification, design, and implementation of the policy. The justificatory politics of REDD+ make powerful use of climate change as a globalizing and totalizing problem for which urgent political solutions must be designed.

In her study of forest conservation and development in Vietnam in chapter 3, Pamela McElwee asks, What are the consequences of using climate change instrumentally to justify this new forest management policy? The practice of elevating carbon management as the primary rationale for REDD+ has managed to avoid other, more locally pressing concerns about uneven land tenure and about the role of ethnic minorities in forest management. "The regulations [about REDD+] that are developing seem to be mostly a rehashing of previously unsuccessful endeavors" and ways of viewing nature and culture that are not really much about climate per se at all (McElwee, p. 96). And as Andrew S. Mathews points out in his Mexican case study in chapter 8, REDD+ demands new flows of information and resources, and these flows can run as much from the periphery to the center as vice versa: "A way of thinking that focuses on the flow of funds *from* the state to the margins systematically obscures flows of information and resources from the periphery *to* the state" (Mathews, p. 217). REDD+ and, by association, climate change opens up new forms of governmentality and bolsters the political authority of the state.

A different example of instrumentalism is offered by Jessica Barnes in chapter 5 in her study of how climate change is framed and understood in Egyptian water politics. Barnes points out the importance of scale and framing in the production of knowledge claims about climate change and water. Her case study also speaks to the entanglement of climate as index and climate as agent to which I have alluded above. But she also asks deeper questions: for whom is climate change useful? "It is also important to ask for whom it [climate change] is perceived as significant and for whom it is not, and to acknowledge the political ramifications of these distinct stances" (Barnes, p. 143). Egyptian farmers, international aid donors, hydrological modelers, government officials, and environmental campaigners all have distinct interests with respect to the framing of climate change in Egypt's hydro-politics. As she concludes, climate change in Egypt may indeed be about water first and foremost, but water in Egypt is not just about climate change.

CLIMATE AND AGENCY

I opened this afterword by suggesting that climate can be read as index or as agent or as both. This designation seems to apply equally powerfully to the idea of anthropogenic climate change. Human-caused climate change is something that can be revealed, for example, by satellite instrumentation or by local observers, and yet something that without the discriminating powers of arcane statistical modeling or the interpretative narratives of sages and prophets can remain hidden (see Rudiak-Gould 2013b, for a development of this idea). And beyond its measurement and indexing, climate change undoubtedly has acquired powerful agency as a cause of outcomes, as an explanation of change in the contemporary world. As new and not-so-new political interests find multiple uses for climate change, our material, social, and imaginative worlds become subject to its powers.

But can or should the study of climate change and human culture be limited to a search for simple categories of cause and effect? What did a changing climate cause in the past? What will a changed climate cause in the future? Is this or that weather extreme caused by human agency? In 2013 in his Holberg Prize lecture, "Which Language Shall We Speak with Gaia?" Bruno Latour asks where agency resides in the world (Latour 2013). He suggests that Lovelock's idea of Gaia—and its inscrutable face—is a way of breaking down the unnatural divide between inanimate matter, which has no agency, and animate human mind, in which all agency resides. Latour claims instead that the political task is to "distribute agency as far and in as differentiated a way as possible" (18) in order to recognize that we are neither pure subjects mastering climate for our pleasure, we might say, nor pure objects being mastered by climate to our detriment. Put differently, and as Dove argues in chapter 1, climate and humans have historically been understood not as two separate domains, one causing or shaping the other. Rather, for much of cultural history and in many places they have been understood to move together, sharing both agency and fate. Perhaps what the idea of anthropogenic climate change has done is to make us (in the enlightened West especially) see more clearly this unavoidable intimacy. It is time we accept it.

This collection of anthropological essays and studies offers an interesting variety of readings, in both contemporary and historical settings, of how this agency has been "distributed" (to use Latour's term) between climate and humans. The essays illustrate very well the versatility of the idea of climate and its changes, both as it has been and as it still is understood across diverse human cultures. There is no one story to tell about climate change. We need a broad variety of insights about climate and its interactions with the human mind and its cultural manifestations; such insights will offer a sufficient number of entry points for human actors to work creatively with the idea of climate change rather than let it paralyze us with fear or fatalism. Neither climate nor humans are fully in charge. The idea of anthropogenic climate change has many uses, and the anthropologists writing in this book have illustrated several of them.

NOTE

1. *"Meeting on UK's run of unusual seasons,"* UK Met Office news blog, June 18, 2013. http://metofficenews.wordpress.com/2013/06/18/meeting-on-uks-run-of-unusual-seasons.

REFERENCES CITED

Anderegg, W., J. Prall, J. Harold, and S. H. Schneider. 2010. "Expert Credibility in Climate Change." *Proceedings of the National Academy of Sciences* 107, no. 27: 12107–9.

Barnes, J., M. Dove, M. Lahsen, A. Mathews, P. McElwee, R. McIntosh, F. Moore, J. O'Reilly, B. Orlove, R. Puri, H. Weiss, and K. Yager. 2013. "Contribution of Anthropology to the Study of Climate Change." *Nature Climate Change* 3, no. 6: 541–44.

Burke, P. 2012. *A Social History of Knowledge.* Volume 2: *From the Encyclopédie to Wikipedia.* Cambridge: Polity Press.

Cook, J., D. Nuccitelli, S. A. Green, M. Richardson, B. Winkler, R. Painting, R. Way, P. Jacobs, and A. Skuce. 2013. "Quantifying the Consensus on Anthropogenic Global Warming in the Scientific Literature." *Environmental Research Letters* 8, 024024.

Corner, A. 2013. "A New Conversation with the Centre-Right About Climate Change: Values, Frames and Narratives." *Climate Outreach and Information Network.* Accessed July 8, 2013. http://www.climateoutreach.org.uk/coin/wp-content/uploads/2013/06/COIN-A-new-conversation-with-the-centre-right-about-climate-change_FINAL-REPORT.pdf.

Donner, S. E. 2007. "Domain of the Gods: An Editorial Essay." *Climatic Change* 85: 231–36.

Doyle, J. 2011. *Mediating Climate Change.* Farnham: Ashgate.

Fleming, J. R., and V. Jankovic. 2011. "Revisiting Klima." *Osiris* 26, no. 1: 1–6.

Henderson-Sellers, A. 2010. "Climatic Change: Communication Changes Over This Journal's First 'Century.'" *Climatic Change* 100: 215–27.

Hulme, M. 2009. *Why We Disagree About Climate Change: Understanding Controversy, Inaction and Opportunity.* Cambridge: Cambridge University Press.

Latour, B. 2013. "Which Language Shall We Speak With Gaia?" Lecture delivered for the Holberg Prize Symposium, Bergen, June 4. Accessed July 9, 2013. http://www.bruno-latour.fr/sites/default/files/128-GAIA-HOLBERG.pdf.

Machin, A. 2013. *Negotiating Climate Change: Radical Democracy and the Illusion of Consensus.* London: Zed Books.

Manzo, K. 2012. "Earthworks: The Geopolitical Visions of Climate Change Cartoons." *Political Geography* 31, no. 8: 481–94.

Rudiak-Gould, P. 2013a. *Climate Change and Tradition in a Small Island State: The Rising Tide.* Abingdon: Routledge.

———. 2013b. "'We Have Seen It With Our Own Eyes': Why We Disagree About Climate Change Visibility." *Weather, Climate and Society* 5, no. 2: 120–32.

Contributors

Jessica Barnes is an Assistant Professor in the Department of Geography and Environment and Sustainability Program at the University of South Carolina. Her research examines the culture and politics of resource use and environmental change in the Middle East. Her first book, *Cultivating the Nile: The Everyday Politics of Water in Egypt*, was published by Duke University Press in 2014, and she has also published articles in *Critique of Anthropology, Geoforum, Social Studies of Science, Geopolitics*, and *Nature Climate Change*. In 2013 Barnes received the Junior Scholar Award from the Anthropology and Environment Society of the American Anthropological Association. Barnes received her doctoral degree from Columbia University in 2010 and in 2011–13 was a postdoctoral associate at the Yale Climate and Energy Institute.

Michael R. Dove is the Margaret K. Musser Professor of Social Ecology, Yale School of Forestry and Environmental Studies; Curator, Peabody Museum of Natural History; Fellow, Whitney Humanities Center; Professor, Department of Anthropology Yale University; and co-coordinator of the joint doctoral program in anthropology and environmental studies at Yale University. His most recent books are *The Banana Tree at the Gate: The History of Marginal Peoples and Global Markets in Borneo* (Yale University Press, 2011, winner of the Julian Steward Award in 2011), *Complicating Conservation: Beyond the Sacred Forest*, coedited with P. E. Sajise and A. Doolittle (Duke University Press, 2011), and *The Anthropology of Climate Change: A Historical Reader* (Wiley/Blackwell, 2014). He was involved in establishing the Climate and Energy Institute at Yale University. His research, teaching, and advising, at the graduate level in the School of Forestry and Environmental Studies and at the undergraduate level in the Environmental Studies major, increasingly focus on disasters and climate change. Dove holds a doctoral degree in anthropology from Stanford University.

Austin Becker is an Assistant Professor of Coastal Planning, Policy, and Design at the University of Rhode Island. His research explores decision-making systems for infrastructure and how to increase resilience to natural disasters induced by climate change. He has published articles in *Climatic Change, Progress in Planning, Journal of Coastal Management* and others. He holds a doctoral degree in Environment and Resources from Stanford University.

Alessandra Giannini is a Research Scientist at the International Research Institute for Climate and Society at Columbia University. Her work focuses on drought, desertification, and climate/environmental change in the semiarid tropics. She has published widely, including articles in *Science, Climatic Change,* and *Environmental Research Letters.* Giannini holds a doctoral degree in Earth and Environmental Sciences from the Lamont-Doherty Earth Observatory of Columbia University.

Grete K. Hovelsrud is Research Director at the Nordland Research Institute and Senior Researcher at the Center for International Climate and Environmental Research in Oslo, Norway. Her work focuses on interdisciplinary studies of vulnerability and adaptation to climate change, resilience and adaptive capacity of coupled social-ecological systems, and on the transformation of society in the context of climatic and societal change. She was a lead author and a contributing author to the IPCC Fifth Assessment Report and a lead author in the Arctic Human Development Report and Arctic Resilience Report. She has published extensively on climate change adaptation, including an edited volume, *Community Adaptation and Vulnerability in Arctic Regions,* coedited with B. Smit (Springer Netherlands, 2010) and articles in *Global Environmental Change, Local Environment,* and *Arctic.* Hovelsrud holds a doctoral degree in anthropology from Brandeis University.

Mike Hulme is Professor of Climate and Culture in the Department of Geography at King's College London. His research employs historical, cultural, and scientific lenses to examine the idea of climate change and its deployment in public and political discourse. His most recent books are *Can Science Fix Climate Change? A Case Against Climate Engineering* (Polity Press, 2014), *Exploring Climate Change Through Science and in Society: An Anthology of Mike Hulme's Essays, Interviews and Speeches* (Routledge, 2013), *Why We Disagree About Climate Change* (Cambridge University Press, 2009), and *Making Climate Change Work for Us,* coedited with H. Neufeldt (Cambridge University Press, 2010). Hulme holds a doctoral degree in applied climatology from the University of Wales, Swansea.

Myanna Lahsen is Senior Researcher in the Earth System Science Center of the Brazilian Institute for Space Research, Brazil. Her research focuses on cultural dimensions of the science–policy interface in the United States and Brazil related to global environmental change, including the sociology of knowledge and development politics. Her publications include articles in *Global Environmental Change, Ecology and Society, Climatic Change, Science, Technology* *and Human Values,* and *American Behavioral Scientist* as well as chapters in a number of edited volumes. Lahsen holds a doctoral degree in cultural anthropology from Rice University.

Heather Lazrus is a Project Scientist at the National Center for Atmospheric Research in Boulder, Colorado. Her research explores the cultural mechanisms through which weather and climate risks are perceived, experienced, and addressed. Her findings contribute to improving the utility of weather forecasts and warnings, reducing social vulnerability to atmospheric and related hazards, and understanding community and cultural adaptations to climate change. Her work has been published in several journals, including the *Annual Review of Anthropology, Current Anthropology, Global Environmental Change,* and *Weather, Climate, and Society* as well as in a number of edited volumes. Lazrus holds a doctoral degree in anthropology from the University of Washington.

Justin S. Mankin is a doctoral student in Environment and Resources at Stanford University. His research aims to constrain the uncertainty essential to understanding and responding to climate change's impacts on people. His work focuses on two of the major sources of uncertainty in climate impacts assessments: the chaos of the climate system and the complexity of how people respond to climate stress. He has published articles in *Climate Dynamics, Climatic Change, Foreign Policy,* and *Journal of International Affairs.* Mankin holds a master's degree in environmental science and policy from Columbia University and a master's degree in global politics from the London School of Economics.

Andrew S. Mathews is Associate Professor of Anthropology at the University of California, Santa Cruz and Co-Director of the Science and Justice Working Group at UCSC. Mathews's research focuses on the interactions between the state, forest-dwelling peoples, and forest landscapes in Mexico. He has published articles in *American Anthropologist, Current Anthropology, Human Ecology,* and *Environmental History,* and a book, *Instituting Nature: Authority, Expertise and Power in Mexican Forests* (MIT Press, 2011). He is currently working on climate change and forests in Mexico and Italy, chestnut and pine domestication, and geoengineering and the imagination of the anthropocene. Mathews holds a doctoral degree in forestry and environmental studies and anthropology from Yale University.

Pamela McElwee is Associate Professor in the Department of Human Ecology at Rutgers, The State University of New Jersey. Her research examines human adaptation to global environmental change, with a focus on biodiversity conservation and climate change in Asia. She has published articles in such journals as *Geoforum, Environmental Conservation,* and *Environmental Management* as well as an edited volume, *Gender and Sustainability: Lessons from Latin America and Asia,* coedited with M. Cruz-Torres (University of Arizona Press, 2012). McElwee earned her doctoral degree in forestry and environmental studies and anthropology at Yale University in 2003.

Roderick J. McIntosh is Professor of Anthropology at Yale University and Curator at the Peabody Museum. As an archaeologist, he has worked for over three decades on the study of ancient urbanism and climate change responses in several parts of Africa. Among his recent books are *The Way the Wind Blows: Climate, History, and Human Action* (Columbia University Press, 2000), *Ancient Middle Niger: Urbanism and the Self-Organizing Landscape* (Cambridge University Press, 2005), and *The Search for Takrur: Archaeological Excavations and Reconnaissance Along the Middle Senegal Valley* (Yale University Publications in Anthropology, in press). McIntosh holds a doctoral degree in anthropology from the University of Cambridge.

Frances C. Moore is a doctoral student in Environment and Resources at Stanford University. Her research focuses on improving estimates of the impacts of climate change on agricultural and food security through an improved understanding of how quickly and effectively farmers will adapt. Her work has been published in *Nature Climate Change, Global Environmental Politics* and the *Journal of Science, Technology and Human Values.* Moore holds a master's degree in environmental science from the Yale School of Forestry and Environmental Studies.

Jessica O'Reilly is Assistant Professor of Anthropology in the Sociology Department at the College of Saint Benedict and Saint John's University. Her first research project examined how Antarctic scientists and policy makers participate in environmental management, and she is currently working on how knowledge is formed within the assessment reports of the Intergovernmental Panel on Climate Change. She has published articles in *PoLAR: Political and Legal Anthropology Review* and *Social Studies of Science* and is completing a book entitled *Technocratic Wilderness: An Ethnography of Scientific Expertise and Environmental Governance in Antarctica* (Cornell University Press, forthcoming). O'Reilly received her doctoral degree in anthropology from the University of California, Santa Cruz.

Ben Orlove is Professor of International and Public Affairs at Columbia University, where he also directs the master's program in Climate and Society and serves as Co-Director of the Center for Research on Environmental Decisions. His early work focused on agriculture, fisheries, and rangelands, and more recently he has studied climate change and glacier retreat, with an emphasis on water, natural hazards, and the loss of iconic landscapes. His recent books include *Lines in the Water: Nature and Culture at Lake Titicaca* (University of California Press, 2002) and *Darkening Peaks: Glacier Retreat, Science and Society* (University of California Press, 2008). Orlove holds a doctoral degree in anthropology from the University of California, Berkeley.

Rajindra K. Puri is Senior Lecturer in Environmental Anthropology and Director of the Centre for Biocultural Diversity in the School of Anthropology and Conservation at the University of Kent, United Kingdom. His research examines the ethnobiological knowledge systems of forest-dwelling peoples in South and Southeast Asia and applies ethnobiological methods to problems in conservation, development, and climate change adaptation. He is the author of *Deadly Dances in the Bornean Rainforest* (KITLV, 2005), an editor of *Ethnobotany in the New Europe*, coedited with M. Pardo and A. Pieroni (Berghahn, 2010), and the editor of the forthcoming volume *Urgent Anthropology: Reflections on Engagement with Threatened Indigenous Peoples, Cultures and Languages* (RAI, 2015). Puri received his doctoral degree in anthropology from the University of Hawaii.

Karina Yager is a Research Scientist for SSAI (Science Systems and Applications, Inc.) at the NASA Goddard Space Flight Center. She is also Visiting Professor at SUNY–Stony Brook in the Sustainability Studies Program, and Resident Scientist at Earth Vision Trust. Her current NASA-ROSES research is focused on the impacts of disappearing tropical glaciers on pastoral agriculture in the Andes. Her interdisciplinary publications are found in multiple journals, including *Nature Climate Change, Journal of Climate, Ecología en Bolivia, Italian Journal of Remote Sensing, Mountain Forum Bulletin, Acta Horticulturae, Pirineos: The Journal of Mountain Ecology,* and a number of edited volumes. Her recent books include *El Parque Nacional Sajama y sus plantas* (Fundación PUMA, 2011) and *Getting the Picture: Our Changing Planet* (Earth Vision Trust, 2014). Her doctoral degree is in anthropology from Yale University.

Index

Page numbers in *italics* refer to illustrations.